SpringerBriefs in Applied Sciences and Technology

PoliMI SpringerBriefs

Springer, in cooperation with Politecnico di Milano, publishes the PoliMI SpringerBriefs, concise summaries of cutting-edge research and practical applications across a wide spectrum of fields. Featuring compact volumes of 50 to 125 (150 as a maximum) pages, the series covers a range of contents from professional to academic in the following research areas carried out at Politecnico:

- Aerospace Engineering
- Bioengineering
- Electrical Engineering
- Energy and Nuclear Science and Technology
- Environmental and Infrastructure Engineering
- Industrial Chemistry and Chemical Engineering
- Information Technology
- Management, Economics and Industrial Engineering
- Materials Engineering
- Mathematical Models and Methods in Engineering
- Mechanical Engineering
- Structural Seismic and Geotechnical Engineering
- Built Environment and Construction Engineering
- Physics
- Design and Technologies
- Urban Planning, Design, and Policy.

http://www.polimi.it

Simonetta Armondi · Gloria Pessina

Critical Spatial Thinking and Extreme Events

Tools for Multiple Disciplines

POLITECNICO
MILANO 1863

Springer

Simonetta Armondi
DAStU
Politecnico di Milano
Milan, Italy

Gloria Pessina
DAStU
Politecnico di Milano
Milan, Italy

ISSN 2191-530X ISSN 2191-5318 (electronic)
SpringerBriefs in Applied Sciences and Technology
ISSN 2282-2577 ISSN 2282-2585 (electronic)
PoliMI SpringerBriefs
ISBN 978-3-032-18874-8 ISBN 978-3-032-18875-5 (eBook)
https://doi.org/10.1007/978-3-032-18875-5

This work was supported by Politecnico di Milano.

This Springer imprint is published by the registered company Springer Nature Switzerland AG
The registered company address is: Gewerbestrasse 11, 6330 Cham, Switzerland

Preface

This book explores the complexity of extreme socio-environmental events and the possible ways to cope with them by mobilising critical and spatial thinking. Being set in "critical disaster studies", the book proposes a critique of dominant intellectual traditions and practices. However, unlike other texts based exclusively on critical deconstruction, the book aims at stirring an anticipatory, prefigurative and generative disaster-related sensibility for hope and new preparedness. Therefore, the book proposes to undo catastrophic and routine disaster-related narratives, approaches, and actions often presented as necessary, due to emergency reasons, by referring to the rich debate on these topics developed in urban planning, policy, critical geography, environmental sociology, and political ecology. Technical solutions and new governance arrangements conceived to respond to disasters are closely observed to highlight what is at stake in an emergency, the differentiated impacts of crises and disasters, their temporality and the medium/long-term effects of the response. Grounded in an extensive literature review of the topic, and with relevant case studies in different parts of the world, the book will appeal to all those interested in applying critical thinking to contemporary territories in crisis. In particular, the book addresses mainly the curiosity of researchers and master students from various disciplinary backgrounds, as well as policymakers.

Chapter 1 "Why Critical Spatial Thinking to Understand Extreme Events" (authors: Simonetta Armondi and Gloria Pessina) introduces the background of critical spatial theory, proposing to observe contemporary urgent phenomena such as disasters through an approach that examines dominant intellectual traditions and mainstream responses.

Chapter 2 "Framing Extreme Events in Contemporary Territories" (author: Simonetta Armondi) identifies mainstream approaches and responses to disasters and their ineffectiveness while highlighting emerging perspectives to reframe the topic (through Environmental Justice, Political Ecology, Urban Political Ecology, Decolonial Studies).

Chapter 3 "Ecological Disasters, Violence and Environmental Justice Movements. Case Studies from India, Italy and the US" (author: Gloria Pessina) explores often overlooked forms of disasters, typically not recognised as such because of the lack of sudden traumatic and visible events. In most cases, such disasters have an ecological nature and require a diachronic approach to identify them. Therefore, they are often considered "slow disasters" or manifestations of "slow violence". The chapter proposes a review of responses to such kinds of disasters, mostly referring to environmental justice movements in various parts of the world. Moreover, the chapter presents a selection of case studies from Europe, Asia, and the US, directly observed and analysed by the authors of the proposed monograph.

In Chap. 4 "Crises and Extreme Events as a Portal for Reimagining Other Futures" (author: Simonetta Armondi), after having proposed a wide overview of different forms of extreme events, related frameworks and concepts, and a variety of case studies, the book ends with alternative responses to disasters elaborated in different cultural contexts and by various authors.

Through the four chapters, this book offers tools to understand the complexity of extreme socio-environmental events and to cope with them by mobilising critical and spatial thinking. The book is set in critical disaster studies and therefore proposes a critique of dominant intellectual traditions and practices. However, unlike other texts based exclusively on critical deconstruction, the proposed book aims to cultivate a research sensibility, that is anticipatory, prefigurative, generative and unleashing creative, experimental possibilities for hope, care and new preparedness.

Moreover, the book proposes to unpack catastrophic and routine disasters related narratives, approaches, and actions often presented as necessary due to emergency reasons. Technical solutions and new governance arrangements created to respond to disasters are therefore closely observed to highlight what is at stake in an emergency, the differentiated impact of disasters, their temporality and the medium/long-term effects of the response.

The COVID-19 pandemic added to existing inequalities and imbalances, becoming a significant (planning and policy) disaster for several countries, particularly for some specific territories. In this context, planners and geographers once again reflected on preparedness, uncertainty, risks, territorial fragilities, and the need to promote a more comprehensive and less technocratic approach to disasters, able to combine multiple perspectives and disciplines.

After the pandemic, the two authors decided to start the course on "Strengthening Critical Spatial Thinking" for the Ph.D. School at Politecnico di Milano, our university. This teaching activity has been addressing Ph.D. candidates from various disciplinary backgrounds, including energy, civil, environmental, management engineering, urban planning, architectural design and conservation, and interior and exhibition design. The geographical origins and cultural backgrounds of the students have been equally diverse. Therefore, the course has combined a broad international literature developed in critical disaster studies with the rich debate on disasters developed in urban planning, policy, and geography, with inputs from critical geography, environmental sociology and political ecology.

Consequently, grounded in an extensive literature review of the topic and with relevant case studies in different parts of the world, the book will appeal to policymakers, planners, geographers, and all those interested in applying critical thinking to contemporary territories in crisis. The book particularly targets Ph.D. students,

M.Sc. Students and practitioners who are currently engaged in research on risks, crises, and disasters with a technical or technological approach to them (e.g. environmental, energy, civil, informatics engineers). The book is equally relevant for Ph.D. students, M.Sc. Students and young researchers dealing with contemporary spaces and territories at multiple scales (architects, architectural conservationists, urban planners, urban designers, geographers, and environmental/territorial sociologists). Designers and artists currently engaging with research on and representation of disasters are also considered by the authors.

The book is an invitation not so much to say what it is, or what it should be, but to provoke thought, to "slow down" reasoning, producing a slightly different awareness of the problem. To conclude, the book is an opportunity to give frames and tools of critical thinking and spatial analysis to a broad audience, addressing the topics from a more territorial and less technical perspective while promoting a deeper understanding of the policy processes and narratives involved in disasters and post-disaster recovery, accentuating the role of space—and the different issue mobilized by spaces—at different scales.

Milan, Italy

Simonetta Armondi
Gloria Pessina

Acknowledgments

The publication of the book was supported by CRAFT (Competence centeR Anti-Fragile Territories). CRAFT is an initiative of the Department of Architecture and Urban Studies—Politecnico di Milano—funded under the project Department of Excellence 2023–2027 promoted by the Italian Ministry of University and Research. The authors would also like to thank Franca Dattrino for proofreading the text.

POLITECNICO MILANO 1863 | DEPARTMENT OF ARCHITECTURE AND URBAN STUDIES

craft competence center anti-fragile territories

Contents

Chapter 1
Why Critical Spatial Thinking to Understand Extreme Events

Abstract The chapter introduces the main aims, contents, and structure of the book. By adopting a perspective stemming from critical disaster studies, environmental justice, and political ecology, the chapter clarifies the need for more complex and diachronic understanding of extreme events, going beyond immediate responses and technical solutions. Critical theory, as originated in the fields of geography, urban and disaster studies is therefore a paramount tool to understand and live in contemporary crises. The chapter underlines the fundamental socio-spatial dimensions and effects of disasters, often not considered by more traditional studies on the topic. By briefly introducing case studies that will be extensively developed in the subsequent parts of the book, the chapter anticipates the frequent use of post-disaster responses to consolidate forms of socio-environmental violence and injustice. Eventually, the chapter emphasises the need to decolonise disaster studies by incorporating locally produced forms of knowledge and response to extreme events.

Keywords Critical disaster studies · Environmental justice · Political ecology · Urban political ecology · Disaster capitalism · State violence · Environmental violence · Decolonising disaster

1.1 Aims of the Book

On the night of 29 October 2024, the Valencia region in Spain reportedly received more than 300 l/m^2 of rain, equivalent to the annual rainfall, in just eight hours. This led to unprecedented flooding, resulting in casualties and loss of property. This devastating catastrophe clearly shows that climate governance failure equals climate catastrophe, even in relatively wealthy countries. Even though one of the most quoted phrases in environmental geography, after Hurricane Katrina in New Orleans, is "there's no such thing as a natural disaster" (Smith 2006), the emphasis of media representation is on the quantity of rain and not on the crucial effective preparedness of policy and planning tools at different scales needed to address climate change caused by anthropogenic interventions.

S. Armondi and G. Pessina, *Critical Spatial Thinking and Extreme Events*,
PoliMI SpringerBriefs, https://doi.org/10.1007/978-3-032-18875-5_1

Hazard events—the interaction between a hazard and people, places and the environment, which "requires" a public response—span minor incidents that are managed locally to global incidents that require international responses. Disaster researchers have long worked to categorise hazards, disasters and catastrophic events in several typologies through space and time (Montano and Savitt 2023). As Quarantelli (2005) stressed in the wake of Hurricane Katrina: "The distinction we draw between catastrophes and disasters is not just an academic exercise (…)". What is crucial is that catastrophes require some different kinds of planning and managing as well as even major disasters. According to Barton, in his foundational book, *Communities in Disaster: A Sociological Analysis of Collective Stress Situations*: "The best way to develop understanding of this set of events is to find several basic dimensions on which these situations vary, which will help account for a wide range of differences. Combinations of positions on these dimensions will then constitute types; members of each type will behave similarly, if we have chosen the dimensions well" (Barton 1969, p. 38).

However, beyond disasters per se, they can be assumed to be "crisis events", and not merely natural events that unfold in isolation (Armondi 2025). From a social sciences perspective, floods, earthquakes, droughts, and pollution of water and soil are deeply political and shaped by economic and political actions framed before and after the event. These decisions determine vulnerability, risk, and the capacity for recovery. Scholars such as Oliver-Smith and Hoffman (1999), Pearce (2003), and Perry (2007) have argued that disasters are not just ruptures in the perceived routine of everyday life. They are outcomes of historical, political, and social processes. They disrupt the status quo, revealing the often-hidden forces of power affecting people who suffer, survive, and are left behind.[1]

Thus, it is increasingly well understood that "natural" disasters are a misnomer. What we refer to as natural events—earthquakes, floods, pandemics, hurricanes—are palimpsests for much broader processes, beyond emergency and concerning also social, political and economic phenomena resulting from the interplay of environmental forces and human choices. These events (catastrophic disasters), as well as slow (or routine) disasters related to pollution, are politically constructed, and the way they are experienced, responded to, and recovered from is deeply influenced by power relations and societal structures (Oliver-Smith 1999; Pelling and Dill 2009; Wisner 2001).

Against this backdrop, the context in which policymakers, architects, engineers, geographers, and planners operate today has never been more challenging. Disasters have become paramount tests for government and policymakers. Such context

[1] The Platform "Port Watch" (https://portwatch.imf.org/pages/disruption-monitor), for instance, is a collaborative project between the IMF and the Environmental Change Institute at the University of Oxford. It is a platform that explores how global supply chains are exposed to disruptions in port activity and to assess the domestic and international trade impact of actual and future disasters, such as extreme weather events. Users can simulate the indirect spillover effects of port disruptions to other countries in the maritime trade network. This can be done for both actual and hypothetical port closures. Spillover analysis reveals which countries and sectors are at risk of experiencing trade disruptions, which can help inform timely responses to shocks.

requires a better interdisciplinary understanding of extreme socio-environmental events and related concepts such as risk, uncertainty, environmental justice fragility, and anti-fragility in urban and regional studies, while also analysing case studies of planning tools and policies worldwide.

An essential premise of this book is that we are confronting a situation where restoring lost value is no longer possible. Socio-ecological devastation has reached a critical threshold beyond which no reparative measures can undo the extensive damage to the environment and human societies.

The book is, therefore, a tiny tool to steer one's way through the complexity of extreme events and to try to understand them by mobilising critical thinking on the one hand and spatial thinking on the other. The book is set in "critical disaster studies" main principles (Remes and Horowitz 2021). It stresses reflecting on what it means to be "critical", proposing a critique of dominant intellectual traditions. The way we read about the world and experience it in our everyday lives is only by looking for missing links, changing the visual angle concerning a problem—trying to understand who it was a disaster for and what is at stake in calling it that, and noticing new connections that have been omitted in the past thinking that our understanding of the world and our ability to change it can improve. However, unlike other texts based exclusively on denunciatory deconstruction, the proposed book aims at stirring a research sensibility which, as we see it, is to be anticipatory, prefigurative, generative and unleashing of creative, experimental possibilities for hope and new preparedness.

In this work, we develop and apply a critical geography framework to look at the world through this particular lens, making certain things (e.g. conflicts, powers, new spatialities) stand out and eliciting attention that previously might not have been forthcoming. In this vein, according to Cox (1981), "critical" stands for a paradigm orientation seeking to identify misalignments, incongruencies, and power and spatial inequalities. Robert Cox distinguishes between "problem-solving" and "critical theory". The former aims at helping solve the problems posed within the terms of a particular perspective, which was the starting point. The main strength of this theory is the ability to define parameters to a problem area and "to reduce the statement of a particular problem to a limited number of variables which are amenable to relatively close and precise examination" (Cox 1981, p. 129). It, therefore, "takes the world as it finds it, with the prevailing social and power relationships and the institutions into which they are organised, as the given framework for action" (Cox 1981, p. 128). Consequently, a critical outlook is committed to questioning the backdrop against which, or the framework whereby, problems are identified, and solutions devised, and in so doing, being attentive to the origin and contestation of institutional arrangements, power differentials, agency distribution, knowledge and authority claims, reality definitions, interest and identity attributions.

Furthermore, pivotal to our fields is the capacity to think and engage "spatially". The "spatial" angle of our study implies that material contexts and their multiple scales of planning arrangements and governance structures are essential. This book addresses disasters, risks, and crises related to the spatialities of social-environmental extreme events, unlocking a set of questions aimed at strengthening critical spatial thinking as a paramount tool, remarkably in a polytechnic school.

The book examines, particularly, theories about the taken-for-granted spatiality of extreme events. Indeed, although we often seek to understand one event, we widen the frame to perceive the multi-scalar surrounding territories. In this regard, it is crucial to delve deep into the planetary urbanisation scholarship that re-theorised the environmental impacts of the political economy of territoriality, logistics, state sovereignty, and primary commodity production (Brenner and Schmid 2015; Danyluk 2021; Goh 2021). Additionally, according to Donovan et al. (2024), the critical disaster studies agenda emphasises the significance of positionality on the part of researchers, as well as the prioritisation of local epistemologies (Gaillard 2019, 2021; Goodall et al. 2022) and "minor" approaches reforged in the field of critical geography by Katz (1996, 2017) beyond strictly political economy aspects (see Chaps. 2 and 4).

Several scholars emphasise that disasters cannot be understood apart from the geo-economics and political dynamics that govern them, from the distribution of resources to the priorities of recovery. Indeed, the writings of Grove (2013a, b), Pelling and Dill (2009), Anderson (2016) and Armondi et al. (2023) have moved disaster studies toward a more critical understanding of the ways capitalism, politics, and broader societal forces outline preparedness, emergency tools, disaster risk and resilience.

The study of disaster has evolved significantly since the early twentieth century. Initially focused on the physical determinants of hazards and the management of immediate impact, disaster research has grown to recognise the social dimensions of vulnerability, risk, and recovery. Through the work of scholars like Quarantelli (1999) and Solnit (2009), we have come to understand that how societies respond to disasters is primarily socially constructed. Social, political, and economic features define, including or excluding, who is the most/more vulnerable to disaster and has the resources to recover. This critical perspective has deepened the understanding of disasters as sites of contestation, where the lines between the victims, the responders, and the state are often blurred and mediated by power.

Causes and recovery, however, remain the least studied aspects of disaster research (Hewitt 1983; Quarantelli 1999). While much of the literature focuses on the immediate aftermath—impact assessment, physical reconstruction, and technical rehabilitation processes—there is growing recognition that disaster recovery is not solely a neutral or technical process. As Vale and Campanella (2005) point out, one of the most straightforward ways to recognise critically the power dynamics of recovery is to observe what is rebuilt, where, and for whom. Additionally, disaster recovery can be a tool of political struggle, with the reconstruction priorities revealing more profound social and political inequalities. However, by reproducing and extending the historical environmental conditions in which disasters occur, recovery can also actively worsen the causes and consequences of disasters. As Quarantelli (1999) and Dello Buono (2012) stressed, recovery can sometimes be mishandled or even captured, leading to "second" or "third" disasters—additional forms of harm caused by the mishandling of the recovery process itself. The example of two subsequent devastating floods in 2023 and 2024 in Emilia-Romagna is emblematic in this regard. According to the analysis of the National System for Environmental Protection (SNPA 2023), in one of the most industrialised and polluted Italian territories, soil sealing and overbuilding have altered natural basins and prevented natural infiltration, increasing the intensity and risk of flooding.

To bridge the existing gaps in the disaster research literature, the book makes extensive use of case study research to highlight the diachronic development of selected disasters in different parts of the world (event, response, medium-long term consequences and changes in the power dynamics, social and material legacies, territorial fragilisation processes, new planning arrangements, conflicts etc.), as well as the less visible aspects of phenomena that might not be recognised as disasters at first sight. Such an approach aligns with the aim of the book to incorporate local epistemologies and grounded knowledge production stemming from specific disasters in different social, cultural and political contexts.

1.2 Contents of the Book

This book explores the politicisation of disaster and the critical role of space, power, politics, and capitalism in shaping disaster causes, response and recovery. In recent years, scholars in critical geography, political ecology, environmental humanities and related disciplines have made significant advances in uncovering the political dimensions of disaster. However, as Olson (2008) argues, there remains a reluctance to engage with the full range and scale of the politics of disaster in many disaster studies spheres.

By "critically" interrogating the economic and political forces at play in disaster roots and recovery, this work sheds light on how power is exerted before and in the wake of catastrophe and how these dynamics shape "spatially" the lives of those affected. This perspective is crucial to understand how disasters unfold and how their spatialities and scales are used to reinforce or challenge existing social orders. It is also crucial to unveil the spatial context of disaster: It does locate the socio-ecological dynamics of urban space within a relational set of flows that extends beyond a bounded conceptualisation of urban form (Swyngedouw and Heynen 2003).

This book is designed to assist readers, including practitioners, graduates, and Ph.D. students, in investigating these themes in greater depth. It encourages critical reflection on cultivating hope, creativity, and resistance within research, practice, and everyday life. Through engaging with the work of scholars and activists who advocate for new visions of society, readers will be invited to consider how their own professional or academic endeavours might contribute to the development of more just and sustainable futures.

After an introduction to the notion of critical thinking, the book examines the mainstream perspectives and emerging paradigms in the study of disasters. The text presents established, mainstream approaches to disasters, focusing on immediate managerial emergency response. It introduces new paradigms that challenge and expand our understanding of disaster events (Chap. 2). Notably, it examines critical thinking through theories that mobilise disasters through the lenses of environmental justice, political ecology, and urban political ecology (Gandy 2022), considering on one side how the urbanisation of capital and global processes influence disaster vulnerability and on the other (side) the wrong answers to an incorrectly formulated policy and planning question (Kapucu 2008).

This book (particularly Chap. 2) examines the role of capitalism in shaping disasters, comparing and arguing the notions of the Capitalocene and the Anthropocene (Moore 2016). This is a crucial juxtaposition to focus on the current era and actions in which human activity is causing severe environmental degradation and social disruption. In the book, following the argument posited by critical geographers and environmental politics scholars, risk and disasters are assumed to be socially constructed and shaped by power dynamics, cultural norms, and economic and political decisions rather than being natural or inevitable, quick shock events.

The evidence from around the globe supports the broader claims that recovery's reproductive propensities are linked to shock and "disaster capitalism" (Klein 2007) or "disaster colonialism" (Rivera 2022). On the one hand, Naomi Klein's disaster capitalism concept highlights how crises, including natural disasters, are often exploited for economic gain. After a disaster, influential private corporations and political elites may seize the opportunity to privatise public services, increase military expenditure, or push neoliberal economic reforms that overly benefit the wealthier groups. Instead of being a moment of collective recovery, the reconstruction phase after a disaster may become an opportunity for elites to enhance their power and wealth, further deepening social inequalities.

On the other hand, postcolonial and decolonial scholarships have advanced the argument that both historical and current processes like colonialism, extractivism, and global inequalities—through logistics and manufacturing infrastructures, energy and critical resources exploitation—contribute to societal and spatial vulnerability, framing the coloniality of disasters as outcomes of oppression and developmentalism (Arboleda 2020; Bonilla 2020; Armondi 2021; Armondi and Geroldi 2022; Armondi et al. 2024).

The mobilisation of decolonial scholarship is vital in our book because it allows us to concatenate the theme of extreme events to the issue of climate change. The decolonial option is not an ancillary task, but at a minimum, it entails avoiding reproducing colonial relations as academics. Such an approach requires, firstly, a better understanding of the meanings and practices of disasters and climate (as in the case of disaster, it is increasingly recognised that the ways climate change is known and represented "matters"). Secondly, it needs a focus on "leaving unnoticed" power relations around race, gender and sexuality, spaces, and other intersectional axes and how these have shaped access and control over legitimate knowledge. Thirdly, it also means that decolonial and other non-Eurocentric knowledge take centre stage to reinterpret environmental and epistemic challenges in the face of the nexus between extreme events and climate change (Sultana 2024).

Many case studies demonstrate that the decision to declare a problem a "disaster" or a "crisis" is an act with spatial and economic consequences (Pellizzoni et al. 2022). In this book, we examine the intricate relationship between disasters and common—and contested—concepts such as development (economic growth at all costs) and resilience. We emphasise how disasters are not isolated events but catalysts that reveal and exacerbate existing social, political, and economic weaknesses within societies. The aim is to explore how disasters magnify pre-existing social and spatial fragilities and inequalities, disproportionately impacting marginalised groups, such as low-income communities, coloured people, and women. It highlights how disasters expose the structural flaws within social systems, making clear that the impacts are not merely a consequence of the disaster itself but are deeply entangled with the disparities already present in society.

This book delves into the post-disaster reconstruction process, particularly the narratives that shape how these efforts are represented in the public discourse. It critically examines whether the focus is on addressing the root causes of vulnerability, such as poverty, political marginalisation, and environmental degradation, or if efforts restore the status quo, maintaining the pre-existing power structures that contributed to the impact of the disaster in the first place.

The authors explore disaster response and reconstruction efforts through case studies in Gujarat (India), Sardinia (Italy), and Louisiana (USA) (Chap. 3). These examples, examined through the lens of political ecology, environmental justice, and critical urban studies, emphasise how local contexts—such as political dynamics, economic structures, and environmental factors—shape recovery strategies. They also highlight how global pressures, such as international aid systems and global economic policies, influence the scale and effectiveness of recovery efforts.

The traditional spatial approach in planning and policies to risk often falls into the trap of environmental determinism—the idea that natural hazards alone shape the fate of populations. This view fails to recognise how human settlements, social systems, and power structures influence border-making, "sacrifice zoning", and disaster outcomes, also through forms of environmental violence. According to Pessina (2022a), slow violence is "spectacle deficient" (Nixon 2011, p. 47) and tends to go ignored until its effects start to be perceived through biodiversity loss, pollution (Watts 2015; Zierler 2011) and the occurrence of illness.

The concepts of "slow/soft violence/disasters" are introduced (Chaps. 2 and 3), which are defined as events that unfold over extended periods, such as environmental degradation related to pollution caused by industrial sites (Baruah 2023). By incorporating a critical geography perspective, we can better understand how risk is spatially uneven, with particular emphasis on the role of socio-economic and political factors in shaping exposure to hazard. The book argues that these events should be understood in terms of both space and time, critically exploring through the notion of "toxic heritage" as both a material phenomenon and a concept (Kryder-Reid and May 2024), the path dependence of a slow disaster and the political economy of the specific context.

Disaster recovery holds an ambiguous status in debates on disaster policy, tools, and politics. Whilst some scholars have documented recovery trend to reproduce and exacerbate the historical conditions that underpin disasters and guide their uneven effects, others emphasise its potential to activate attempts to change these conditions and begin new development pathways (Donovan et al. 2024) and possibilities (Amoore 2013). In contemporary scholarship and academia, there is growing recognition of the need to reimagine how we live, work, and relate to one another in a world increasingly shaped by capitalism and neoliberalism. Drawing on the perspectives of Gibson-Graham (2006) and others, this book (in Chap. 4) challenges the over-theorization of capitalism by highlighting the everyday forms of experimentation, creativity, and vision that offer alternative possibilities for societal transformation (Anderson and Fenton 2008; Head 2016; Pickerill and Chatterton 2006).

Disruptions, paradoxically, can generate opportunities for new ways of doing and being and present possibilities for hope and change (Head 2016). Research is ever more acknowledged as a generative and performative practice that contributes to shape the world we come to live in, as Cameron and Hicks (2014) posit. This hope is not an ingenuous positivity; instead, it is a call to action and struggle. It is an invitation to actively work toward creating alternative policy and planning tools beyond capitalism and neoliberalism's constraints.

Furthermore, this section explores the concept of "broken-world thinking" (Jackson 2014), which helps to emphasise understanding and responding to disasters not just as events to be "recovered" from but as moments to rethink and repair infrastructure systems and places. Following O'Grady and Shaw (2023), this section expands the conceptualisation of recovery transformative potential by engaging in discussions around reparations. In doing so, it moves towards "disaster reparations". It focuses on how disaster policy-setting should go beyond traditional models to include, paradoxically, uncertainty (Moroni and Chiffi 2021) and preparedness (Armondi 2023), not just to "bounce back better" but to adapt and thrive in the face of future challenges. According to Armondi (2023), although preparedness has been primarily investigated in crisis and disaster studies (Hémond and Robert 2012), the concept is not yet a mainstream narrative for urban policy and planning research. Formulated by national institutions and driven forward by skilled professionals, preparedness is empirically rich but theoretically weak on the macro scale (Staupe-Delgado and Kruke 2018).

Preparedness can have greater analytical depth by elaborating on distinct options for decision-making and complexity, and uncertainty management in cities and regions. Whilst definitions of urban preparedness for crises and shock impacts may differ across academic and policy debates, resilience is likely a dimension that comes first. As opposed to depending on reactive measures enacted in response to disasters, the preparedness approach could create an interpretation of complexity and uncertainty that underlines the need to break up existing policy silos, combat fragmentation and establish multi-level, multi-sector and cross-territorial working relationships.

Finally, the book, according to the "broken-world thinking" (Jackson 2014) highlights the role of hope (Cretney 2017), improvisation (Kendra and Wachtendorf 2006; Ravazzi 2024) and care (Puig de la Bellacasa 2011, 2022) both in disaster preparedness and planning and policy response, encouraging flexibility and anti-fragility and using unconventional and unexpected (spatial) answers in moments of crisis when plans and projects often fail (Chap. 4). Here, the focus lies on exploring the hopes and possibilities for ecological reparation, pointing to the fragile everyday materialities that connect humans and non-humans. Focusing on minor practices of care and repair shifts attention away from large-scale environmental remediation projects, which aim at addressing ecosystem damages that may have wide-reaching and unpredictable consequences for both proximate and remote ecological relations.

1.3 Observing Disasters Through Space and Time, Starting from Italy and for a Polytechnic Readership

In Italy, particularly after the 2009 earthquake in L'Aquila, the topic of disasters gained increasing attention among the public and in the academic debate of several disciplines, including geography, urban planning, heritage and conservation studies, civil and environmental engineering, as well as environmental and territorial sociology (Pessina 2022b). In the following years, further destructive events took place in the country, including the earthquakes in the Emilia-Romagna region (2012) and in Central Italy (2016–2017), and the floods in Genoa (2010, 2014), Massa Carrara (2012), Sardinia (2013), Marche (2014), Northern Milan (2014), Calabria Ionica (2015), Livorno (2017), Liguria, Piedmont, and Venice (2019). In 2018, a major windstorm (Vaia) hit the northeastern Alps, while summer fires have been increasing their frequency, particularly in the Southern regions of Sicily, Campania, Calabria, and Sardinia.

Hence, urban planners in Italy have been increasingly considering the country as characterised by major fragilities that needed to be analysed with appropriate methods and addressed by adequate policies and plans (Pileri 2019). It is in this context that the research project "Territorial Fragilities" (TF) of the Department of Architecture and Urban Studies (DAStU) of Politecnico di Milano took place after DAStU was awarded as one of the recipients of the Departments of Excellence 2018–2022 special fund by the Italian Ministry of University and Research (Balducci 2019). Even though the topic of disasters was not at the core of the project (Dezio et al. 2019), it has been widely present in the works of several Italian architects and planners involved in various research activities related to it, such as *Ricomporre i divari* (Coppola et al. 2021a), a project aimed at identifying and addressing existing territorial imbalances in Italy. In this framework, the damage to the territory following major disruptive events was associated with a variety of already existing elements of concern, including soil sealing, overbuilding, mismanagement of forest and mountain areas, as well as polluting land uses (Pileri 2019; Becciu et al. 2021; Carrosio and De Renzis 2021; Curci et al. 2021; Magoni et al. 2021; Lanzani et al. 2021). As the editors of the volume resulting from the *Ricomporre i divari* research highlighted, urban and territorial planning in Italy have not been radically reformed in the light of disaster prevention and adaptation strategies (Coppola et al. 2021b), despite the widespread acknowledgement of the environmental footprint of human activities on climate change and the presence of numerous studies also in the planning domain. Especially after the L'Aquila earthquake in 2009, planners have been engaging with the topics of risks and disasters (e.g. Angrilli 2013; Menoni 2017) mostly through the development of adaptation strategies and policy recommendations.

As research works in the fields of critical urban studies, geography and territorial sociology on the earthquake in L'Aquila (Coppola et al. 2018; Valent 2019), in Central Italy (Mela et al. 2017; Olori 2024) and in Irpinia (Zaccaria 2023) have shown, the response to such events has not been adequate and has generated further disasters for the local populations and conflicts among the involved actors. While economic compensations have been prioritised, newly created technical and operational structures such as "Casa Italia" and "Italia Sicura" still approached risks and disasters as emergencies (Balducci et al. 2021), contributing to the consolidation of inequalities and power unbalances in the affected territories (Coppola and Pessina 2025). Therefore, a long-standing tradition of studies on environmental conflicts in Italy (e.g. Bobbio 1994; Lewanski 1996; Lewanski and Liberatore 2002; Pacchi 1999) could once again be mobilised to observe more recent disasters, also in dialogue with the Italian public action sociology (Centemeri 2006), political ecology (Armiero and De Rosa 2016), studies on the contentious role of experts (Pellizzoni 2011) and on the relations between disasters, long-term pollution phenomena, environmental justice, capitalism and climate change (Nebbia 2020).

In Italy, the COVID-19 pandemic in 2020/2021 added to existing inequalities and unbalances turning into a major (planning) disaster for the country, and particularly for some specific territories. In this context, planners once again reflected on uncertainty, risks, territorial fragilities, and the need to promote a more comprehensive and less technocratic approach to disasters, able to combine multiple perspectives and disciplines (Balducci et al. 2020; Curci and Pasqui 2022; Franz 2022).

It is in the aftermath of the pandemic and in the context of the *Territorial Fragilities* project at DAStU—Politecnico di Milano that the two authors of the present book decided to start the course "Strengthening Critical Spatial Thinking" for the Ph.D. School of PoliMi. This teaching activity, which started in the 2022/2023 academic year and is still ongoing, has been addressing Ph.D. candidates from multiple disciplinary backgrounds, including energy, civil, environmental, informatics, management engineering, physics, urban planning, architectural design and conservation, interior and exhibition design. The geographical origins and cultural backgrounds of the students have been equally diverse. Therefore, the course has combined a broad international literature developed in the fields of critical disaster studies, urban studies, political ecology, urban political ecology, and environmental justice with the very rich Italian debate—yet not always critical—on disasters that emerged in the past 15 years in the field of urban planning, with inputs from economic-political geography and environmental sociology. The aim has been to share tools for critical thinking and spatial analysis with the Ph.D. candidates with a more technical and less territorial perspective, while promoting a deeper understanding of the technical processes and narratives involved in disasters and post-disaster recovery for those who were more familiar with space on different scales.

The book *Critical Spatial Thinking and Extreme Events. Tools for Multiple Disciplines* hence elaborates on two years of Ph.D. course and addresses a polytechnic readership active in the academia and beyond. The series Springer Briefs in Applied Sciences and Technology (PoliMi Springer Briefs) has hosted several titles related to disasters, resilience, fragility, and risks, particularly after 2020. Even though societal resilience to disasters concerning transport systems (Atun 2014) and uncertainty in large engineering projects (Caron 2013) had already appeared among the early titles included in PoliMi Springer Briefs, it is with the book *Risk and Resilience. Socio-Spatial and Environmental Challenges* (Balducci et al. 2020) that the series started engaging with more continuity and urgency with the topic. The book originated from the Ph.D. course "Approaches to resilience" at Politecnico di Milano and was promoted within the activities of the research project *Territorial Fragilities* at the DAStU (Department of Excellence 2018–2022). The book edited by Balducci, Chiffi, and Curci is a collection of chapters by multiple authors (philosophers, planners, economists, territorial sociologists). It focuses on the clarification of the concepts of risk and resilience, as well as on the analysis and promotion of new resilient planning approaches.

Coping with the Pandemic in Fragile Cities (Pasqui 2022) further elaborates on the findings of the *Territorial Fragilities* research project and builds on the knowledge proposed by Balducci et al. (2020), while focusing on the difference between the concept of risk and that of uncertainty, in the aftermath of the COVID-19 pandemic. Through the various chapters, the author unpacks/undoes some crucial dimensions of radical uncertainty, adopting a pragmatist and radically possibilist approach. Moreover, the book deals with the concept of preparedness, given its increasing relevance in the public debate on uncertainty, risks, and disasters. Ultimately, by referring to a wider international debate, Pasqui discusses the concept of "antifragility" for new planning practices able to deal with radical uncertainty.

The Polimi Springer Briefs series also published *Enabling Adaptive Water Management to Face Drought Risk in a Changing Climate* (Minucci 2021), a rich and critical review on the adaptive capacity of organisations and institutions, based on an in-depth analysis of specific case studies observed by the author during his Ph.D. stay in Bolivia. Lately, the book series has been hosting a contribution on *Slow Onset Disasters* (Salvalai et al. 2024), in which the authors focus on the urban built environment and on selected public spaces as case studies to develop a framework to assess and mitigate risks centred on the experience of the users. The recent book *Climate Change and Human Systems* (Bernardini 2024) focuses on the barriers to climate adaptation in urban contexts and on the possibilities to overcome them through a reorganisation of urban services according to the principles of flexibility, reversibility, scalability and cost-effectiveness. Eventually, the debate on the topics of risks and adaptation in the book series evolved, leading to a new contribution with *Climate Change, Resilience and Cultural Heritage* (Rajabi 2025). In the book, the author moves between international debates on climate, disaster management, risk and heritage, highlighting the cultural dimension of climate change and its significance for heritage studies. Beyond the theoretical debate that crosses multiple disciplines, the book proposes some exemplary case studies. Considering the variety and the richness of the contributions related to risks, disasters, resilience and climate change in the series, this book, *Critical Spatial Thinking and Extreme Events. Tools for multiple disciplines* shares the concern for a territorial approach to such topics already made explicit by Balducci et al. (2020) and Pasqui (2022). Some of the concepts presented in the two books are further developed in the light of critical disaster studies and critical spatial thinking literature, highlighting the narratives and the power dynamics behind them. The book proposes to cross multiple geographical scales, ranging from global connections to regional and urban dynamics to the impact of disaster on specific built environments and public spaces, which can also become sites for critical imagination. By considering the relevance of the cultural dimension and the possible connections between apparently distant disciplines such as heritage studies and climate science, the book by Rajabi (2025) opens a new and interesting path in the PoliMi Springer Briefs. Such dialogue/relation between social and technical sciences is very strongly present in *Critical Spatial Thinking and Extreme Events. Tools for multiple disciplines.* Eventually, the book engages with Minucci (2021) for the choice to devote part of the text to an in-depth analysis of specific contexts observed through extensive fieldwork and on-site research during a Ph.D. or postdoc research period.

As already tested with the participants in the Ph.D. course "Strengthening Critical Spatial Thinking", case studies allow a highly diverse audience to be engaged with key theoretical concepts and approaches. In the book, case studies will be used in two different ways. On the one hand, widely studied examples of disasters and the related processes (e.g., the L'Aquila earthquake in Italy; Hurricane Katrina in New Orleans, USA) are recalled through the work of international scholars throughout the book for different purposes. First, for having been landmark disasters that influenced and accelerated the national and/or international debate and policies on this topic. Secondly, because the distance intercurred from the disruptive event allowed scholars to observe the evolution of the process through time, as well as its differentiated impact on space. As the American historian Scott Gabriel Knowles declared in an interview, "[…] event thinking [is] looking at a disaster as a discreet event with a beginning, a middle, and an end, something that happened to us. [We would] rather [need] to move our thinking into […] a slow disaster thinking" (Knowles 2017). On the other hand, the book presents a selection of case studies directly investigated by its authors through fieldwork, interviews, data collection, and observation. As Yin puts it, "a case-study is an empirical inquiry that investigates a contemporary phenomenon in depth and within its real-life context" (Yin 2009, p. 18). This second typology of case studies shares "environmental violence" (Pessina 2022a) as a common framework. Two dimensions of this concept are explored through specific cases: on the one hand, infrastructural violence as a form of disaster that cannot be directly linked to a single "natural" event and that can be familiar to a polytechnic readership (engineers, planners, architects); on the other hand, slow violence as a generator of disasters that "occur gradually and out of sight" (Nixon 2011) and hence need to be brought to the attention of the readers.

Infrastructural violence is explored in the book through the investigation of large-scale infrastructure projects to counter the excess or the scarcity of resources (e.g. water), contributing to the structural marginalisation of selected populations and to the irreversible deterioration of ecology (Luxion 2017; Swyngedouw 2015) in the ongoing context of climate change. Moreover, by referring to feminist decolonial literature (Rivera Cusicanqui 1997), infrastructural violence is also interpreted as a product of patriarchal culture, colonialism, and exploitation.

Based on the Ph.D. research conducted by one of the two authors of this book (Pessina 2018), the presented case study is an investigation of major water-related infrastructures in Western India, observed on multiple scales, ranging from the site of a riverfront redevelopment project (Sabarmati Riverfront Development Project) on the urban scale (city of Ahmedabad), to the interstate development of a major irrigation canal (Narmada canal) later turned into the backbone for the industrial and urban development of the state of Gujarat. The case is explored through a political ecology and environmental justice approach, identifying the different phases of the considered projects, the actors involved, the power imbalances and the dominant narratives that turned an infrastructural promise of wealth into a major socio-environmental disaster.

The second case study, based on extensive research conducted by one of the authors (Pessina 2023) in the framework of the *Territorial Fragilities* project (DAStU, Politecnico di Milano), is chosen to investigate in-depth the concept of slow violence and contamination. Located in Italy, in the Southwestern tip of the Sardinia Island, the Portovesme aluminum production pole, in crisis since 2009, has been the result of a sequence of exogenous development policies and choices. Once the port of the Sulcis coal mining area, it became the target of a major energy project aimed a fuelling a major industrial development in the area. While the productive activities had a limited duration, their environmental effects are lasting. Nowadays, the site is among the only two sites in Italy selected for the EU Just Transition Programme (together with Taranto, in the Apulia region) due to its excessive levels of contamination and unemployment.

The case study is observed through a territorial and policy analysis approach, combined with suggestions from the environmental history debate on former critical industrial areas in crisis (Romeo 2019). Therefore, the material dimensions and the toxic legacies (Brenu et al. 2024) of previous development policies and projects related to mining, energy, and aluminium production are investigated, while the main actors are identified. An ethnographic observation complements the study, aiming at showing the differentiated impact of the ongoing energy transition on the different workers in the area.

Such case study is then related a further case of "sacrifice zones" (Lerner 2010) observed by one of the authors through a short visiting period to the Cancer Alley in Louisiana, north of New Orleans. For this case, attention is paid to forms of resistance, conflict and bottom-up knowledge production that involve artistic representations.

1.4 What This Book Is Not About

The book uses a methodology other than the likely more common quantitative approach because it fosters a conceptual and multidisciplinary understanding of how spatial thinking can be applied to extreme events across various disciplines. The book takes into consideration the positionalities of the two authors. However, it mobilises qualitative, critical, and theoretical approaches, mainly in social sciences, to explore extreme events' complexities, backlash and nuances rather than relying on numerical data and statistical models typically associated with quantitative research. Qualitative methods—such as case studies, spatial analysis, and theoretical frameworks—can provide deeper insights into how extreme events are experienced, understood, carved out, and managed across diverse contexts, which might not be easily captured by quantitative measures alone. The book is likely rooted in critical spatial theory (Brenner 2009) and emphasises understanding the social, cultural, economic and political dimensions of space. This approach tends to be less technocentric and more interpretive and reflective, challenging conventional views of space, scale, and event. Nonetheless, the case studies accounts are not presented as standing separate from extreme events, politics and economics, but to understand local, lived experiences as intrinsically woven with political and economic dynamics.

To conclude, the absence of quantitative methodology in the book reflects its goal of engaging with the situated, relational dimensions of spatial knowledge about extreme events. Qualitative methods offer a more nuanced and relational approach to the complexities of the subject matter, which is very significant in a polytechnic book series.

References

Amoore L (2013) The politics of possibility: risk and security beyond probability. Duke University Press, Durham

Anderson B (2016) Governing emergencies: the politics of delay and the logic of response. Trans Inst Br Geogr 41(1):14–26

Anderson B, Fenton J (2008) Editorial introduction: spaces of hope. Space Cult 11(2):76–80

Angrilli M (ed) (2013) L'urbanistica che cambia. Rischi e valori. FrancoAngeli, Milano

Arboleda M (2020) Planetary mine: territories of extraction under late capitalism. Verso Books, Brooklyn, New York

Armiero M, De Rosa SP (2016) Political effluvia: smells, revelations, and the politicization of daily experience in Naples, Italy, in J Thorpe, S Rutherford and LA Sandberg (eds), Methodological Challenges in Nature-Culture and Environmental History Research. Routledge, London, pp. 193–206

Armondi S (2021) Ripensare le politiche energetiche tra emergenze e transizioni. Una prospettiva geografica. Semest Studi Ric Geogr 33(2):7–19

Armondi S (2023) Spatial strategies for getting ready? The weak linkage amid preparedness and urban policy. In: Armondi S, Balducci A, Bovo M, Galimberti B (eds) Cities learning from a pandemic. Towards preparedness. Routledge, London and New York, pp 47–62

Armondi S (2025) Geografie delle crisi. Dati, logistica e urbanizzazione del mondo. Mimesis, Milano

Armondi S, Geroldi C (2022) Spazi della logistica e paesaggio del petrolio nel Nord Italia. Poteri, relazioni, strategie. In: Fior M, Kërçuku A, Mattioli C, Pessina G (eds) Fragilità nei territori della produzione. FrancoAngeli, Milano, pp 181–198

Armondi S, Balducci A, Bovo M, Galimberti B (eds) (2023) Cities learning from a pandemic: towards preparedness. Routledge, London and New York

Armondi S, Di Vita S, Galimberti B (eds) (2024) Geografie operazionali nel Nord Italia. Figurazioni e strategie spaziali. FrancoAngeli, Milano, pp 29–50
Balducci A, Chiffi D, Curci F (eds) (2020) Risk and resilience. Socio-spatial and environmental challenges. In: Polimi Springer briefs in applied science and technology. Springer, Cham
Balducci A, Caramaschi S, Coppola A, Curci F, Di Giovanni G, di Venosa M, Fontana C, Franz G, Gritti A (2021) Nei territori sismici: principi fondamentali per una legge quadro per le ricostruzioni. In: Coppola A, Del Fabbro M, Lanzani A, Pessina G, Zanfi F (eds) Ricomporre i divari. Politiche e progetti territoriali contro le disuguaglianze e per la transizione ecologica. Il Mulino, Bologna, pp 57–69
Barton AH (1969) Communities in disaster: a sociological analysis of collective stress situations. Doubleday, Garden City, NY
Baruah M (2023) Slow disaster political ecology of hazards and everyday life in the Brahmaputra Valley, Assam. Routledge, London, New York
Becciu G, Lanzani A, Zanfi F (2021) Negli ambiti fluviali: limitazione del rischio idraulico e riequilibrio ambientale e insediativo. In: Coppola A, Del Fabbro M, Lanzani A, Pessina G, Zanfi F (eds) Ricomporre i divari. Politiche e progetti territoriali contro le disuguaglianze e per la transizione ecologica. Il Mulino, Bologna, pp 83–93
Bernardini C (2024) Climate change and human systems. innovative adaptation services as a soft-resilience strategy to tackle emerging risks. In: Polimi Springer briefs in applied science and technology. Springer, Cham
Bobbio L (ed) (1994) Di questo accordo lieto. Sulla risoluzione negoziale dei conflitti ambientali, Rosenberg and Sellier, Torino
Bonilla Y (2020) The coloniality of disaster: race, empire, and the temporal logics of emergency in Puerto Rico, USA. Polit Geogr 78
Brenner N (2009) What is critical urban theory? City 13(2):198–207
Brenner N, Schmid C (2015) Towards a new epistemology of the urban? City 19(2-3) 151–182 https://doi.org/10.1080/13604813.2015.1014712
Brenu N, Pessina G, Tiganea O (eds) (2024) Eredità tossiche/toxic legacies. Special Issue, Officina
Balducci A (2019) Dobbiamo occuparci di fragilità territoriali. Urban@it (blog), 17 Apr 2019. Available at: https://www.urbanit.it/dobbiamo-occuparci-di-fragilita-territoriali/
Cameron J, Hicks J (2014) Performative research for a climate politics of hope: rethinking geographic scale, “impact” scale, and markets: performative research for a climate politics of hope. Antipode 46(1):53–71
Caron F (2013) Managing the Continuum: Certainty Uncertainty Unpredictability in Large Engineering Projects Springer Milan Milano
Carrosio G, De Renzis A (2021) Nelle aree interne: una corretta gestione e valorizzazione del capitale naturale. In: Coppola A, Del Fabbro M, Lanzani A, Pessina G, Zanfi F (eds) Ricomporre i divari. Politiche e progetti territoriali contro le disuguaglianze e per la transizione ecologica. Il Mulino, Bologna, pp 47–56
Centemeri L (2006) Ritorno a Seveso. Il danno ambientale, il suo riconoscimento, la sua riparazione. Bruno Mondadori, Milano
Coppola A, Pessina G (2025) Le crisi territoriali tra mutamento dell’intervento statale, nuove strategie assicurative e nuove disuguaglianze. Prospettive dagli Stati Uniti e dall’Italia. In: Perocco F, Pirina G (eds) Le disuguaglianze territoriali in Italia. Cause, forme, conseguenze. Edizioni Ca’ Foscari, Venezia, pp. 335–358
Coppola A, Fontana C, Gingardi V (2018) Envisaging L’Aquila. Strategies, spatialities and sociabilities of a recovering city. Professional Dreamers
Coppola A, Del Fabbro M, Lanzani A, Pessina G, Zanfi F (eds) (2021a) Ricomporre i divari. Politiche e progetti territoriali contro le disuguaglianze e per la transizione ecologica. Il Mulino, Bologna
Coppola A, Lanzani A, Zanfi F (2021b) Tra Eredità, Riscoperte e un Futuro Diverso: Ripensare le Politiche Urbanistiche e Territoriali. In: Coppola A, Del Fabbro M, Lanzani A, Pessina G,

Zanfi F (eds) Ricomporre i divari. Politiche e progetti territoriali contro le disuguaglianze e per la transizione ecologica. Il Mulino, Bologna, pp 13–37
Cox R (1981) Social forces, states and world orders: beyond international relations theory. Millennium J Int Stud 10(2):126–55
Cretney RM (2017) Towards a critical geography of disaster recovery politics: perspectives on crisis and hope. Geogr Compass 11(1)
Curci F, Pasqui G (2022) I territori fragili della pandemia: interpretazioni, luoghi, progetti, politiche. Territorio 97. Available at: https://journals.francoangeli.it/index.php/territorioOA/issue/view/1000
Curci F, Mininni M, Nanni G, Zanchini E, Zanfi F (2021) Nelle aree costiere: accessibilità, sicurezza e risposta al cambiamento climatico. In: Coppola A, Del Fabbro M, Lanzani A, Pessina G, Zanfi F (eds) Ricomporre i divari. Politiche e progetti territoriali contro le disuguaglianze e per la transizione ecologica. Il Mulino, Bologna, pp 71–82
Danyluk M (2021) Supply-chain urbanism: constructing and contesting the logistics city. Ann Am Assoc Geogr 111(7):2149–2164
Dello Buono R (2012) Writing the relationship of resistance and social change. In: Fasenfest D, Dello Buono R (eds) Social change, resistance and social practices: studies in critical social sciences. Haymarket, Chicago, IL, London, pp 1–55
Dezio C et al (2019) Territorial fragilities in Italy. Defining a common lexicon. Territorio 91:22–54
Donovan A, Hall TE, Morin J, Smith C, Walshe R (2024) Critical geographies of disaster and the geographical imagination. Prog Environ Geogr 3(3):212–230
Franz G (2022) L'umanità a un bivio. Il dilemma della sostenibilità a trent'anni da Rio de Janeiro. Mimesis, Milano
Atun F (2014) Improving Societal Resilience to Disasters A Case Study of London's Transportation System Springer International Publishing Cham
Gaillard J (2019) Disaster studies inside out. Disasters 43:S7–S17
Gaillard JC (2021) The invention of disaster: power and knowledge in discourses on hazard and vulnerability. Routledge, London
Gandy M (2022) Urban political ecology: a critical reconfiguration. Prog Hum Geogr 46(1):21–43
Gibson-Graham JK (2006) A postcapitalist politics. University of Minnesota Press, Minneapolis
Goh K (2021) Form and flow: the spatial politics of urban resilience and climate justice. MIT Press, Cambridge
Goodall S, Khalid Z, Del Pinto M (2022) Disaster conversations: intersecting perspectives on cross-cultural disaster research. Disaster Prev Manag Int J 31:10–21
Grove KJ (2013a) On resilience politics: from transformation to subversion. Resilience 1(2):146–153
Grove KJ (2013b) From emergency management to managing emergence: a genealogy of disaster management in Jamaica. Ann Assoc Am Geogr 103(3):570–588
Head L (2016) Hope and grief in the Anthropocene. Routledge, London
Hémond Y, Robert B (2012) Preparedness: the state of the art and future prospects. Disaster Prev Manag 21(4):404–417
Hewitt K (ed) (1983) Interpretations of calamity. Allen and Unwin, Boston
Jackson SD (2014) Rethinking repair. In: Gillespie T, Boczkowski PJ, Foot KA (eds) Inside technology. Media technologies: essays on communication, materiality, and society. MIT Press, Cambridge, MA, London
Kapucu N (2008) Planning for disasters and responding to catastrophes: error of the third type in disaster policy and planning. Int J Public Policy 3(5/6):313–327
Katz C (1996) Towards minor theory. Environ Plan D Soc Space 14:487–499
Katz C (2017) Revisiting minor theory. Environ Plan D Soc Space 35:596–599
Kendra J, Wachtendorf T (2006) Improvisation, creativity, and the art of emergency management. Preliminary paper 357, University of Delaware, Disasters Research Center. http://udspace.udel.edu/handle/19716/3054
Klein N (2007) The shock doctrine: the rise of disaster capitalism. Metropolitan Books, New York

Knowles SG (2017) A scholar looks at history to chart a new narrative on natural disasters. Interview on WNYC, 13 Sept 2017. Available at: https://www.wnyc.org/story/scholar-looks-history-chart-narrative-natural-disasters/
Kryder-Reid E, May S (2024) Toxic heritage legacies, futures, and environmental injustice. Routledge, New York
Lanzani A, De Leo D, Mattioli C, Morello E, Zanfi F (2021) Nell'Italia di Mezzo: rigenerazione e valorizzazione dei territori della produzione. In: Coppola A, Del Fabbro M, Lanzani A, Pessina G, Zanfi F (eds) Ricomporre i divari. Politiche e progetti territoriali contro le disuguaglianze e per la transizione ecologica. Il Mulino, Bologna, pp 107–115
Lerner S (2010) Sacrifice zones: the front lines of toxic chemical exposure in the United States. MIT Press, Cambridge, MA
Lewanski R (1996) Environmental dispute resolution in Italy. In: Weidner H (ed) Alternative dispute resolution in environmental conflicts. Experiences in 12 countries. Sigma, Berlin
Lewanski R, Liberatore A (2002) Environmental protection in Italy: analysing the local, national and European community levels of policy making. In: Desay U (ed) Environmental politics and policy in industrialised countries. MIT Press, Cambridge, MA, London, UK
Luxion M (2017) Nation-building, industrialisation, and spectacle. Political functions of Gujarat's Narmada pipeline project. Water Altern 10(2):208–232
Magoni M, Pessina G, Radaelli R (2021) Nel Bacino Padano: una strategia di disinquinamento dell'aria. In: Coppola A, Del Fabbro M, Lanzani A, Pessina G, Zanfi F (eds) Ricomporre i divari. Politiche e progetti territoriali contro le disuguaglianze e per la transizione ecologica. Il Mulino, Bologna, pp 95–106
Mela A, Mugnano S, Olori D (eds) (2017) Territori vulnerabili. Verso una sociologia dei disastri italiana. FrancoAngeli, Milano
Menoni S (2017) Ricostruzione post-sisma nel Centro Italia: la sintesi urbanistica tra esigenze sociali, aspetti ambientali e fisici, la prevenzione dei rischi. Urban Inf 272:763–767
Minucci G (2021) Enabling adaptive water management to face drought risks in a changing climate. In: Polimi Springer briefs in applied science and technology. Springer, Cham
Montano S, Savitt A (2023) Revisiting emergencies, disasters, and catastrophes: adding duration to the hazard event classification. Int J Mass Emerg Disasters 41(2–3):259–278. https://doi.org/10.1177/02807270231211831
Moore JW (ed) (2016) Anthropocene or Capitalocene? Nature, history, and the crisis of capitalism. PM Press, Oakland
Moroni S, Chiffi D (2021) Complexity and uncertainty: implications for urban planning. In: Portugali J (ed) Handbook on cities and complexity. Edward Elgar Publishing, Cheltenham
Nebbia G (2020) La terra brucia. Per una critica ecologica del capitalismo. Jaca Book, Milano
Nixon R (2011) Slow violence and the environmentalism of the poor. Harvard University Press, Cambridge, MA, London
O'Grady N, Shaw D (2023) Disaster reparations? Rethinking disaster recovery through the politics of affect. Geogr J 189:514–525
Oliver-Smith A (1999) Theorizing disasters: nature, power and culture. In: Oliver-Smith A, Hoffman S (eds) The angry earth: disaster in anthropological perspective. Routledge, New York, pp 23–47
Oliver-Smith A, Hoffman S (eds) (1999) The angry earth: disaster in anthropological perspective. Routledge, New York
Olori D (2024) Il Futuro non è Scritto. Disastro, Territorio e Organizzazione Sociale. Orthotes, Napoli
Olson R (2008) Towards a politics of disaster: losses, values, agendas, and blame. In: Boin A (ed) Crisis management, vol II. Sage, Los Angeles, pp 154–170
Pacchi C (1999) Una via italiana alla gestione dei conflitti ambientali? Equilibri 3:335–342
Pasqui G (2022) Coping with the pandemic in fragile cities. In: Polimi Springer briefs in applied science and technology. Springer, Cham
Pearce L (2003) Disaster management and community planning, and public participation: how to achieve sustainable hazard mitigation. Nat Hazards 28(2–3):211–228

Pelling M, Dill K (2009) Disaster politics: tipping points for change in the adaptation of sociopolitical regimes. Prog Hum Geogr 34(1):21–37
Pellizzoni L (2011) Conflitti ambientali. Esperti, politiche, istituzioni nelle controversie ecologiche. Il Mulino, Bologna
Pellizzoni L, Leonardi E, Asara V (eds) (2022) Handbook of critical environmental politics. Edward Elgar, Cheltenham
Perry RW (2007) What is a disaster? In: Rodrıguez H, Quarantelli EL, Dynes RR (eds) Handbook of disaster research. Springer, New York, pp 1–15
Pessina G (2018) The 'missing conflict' of the Sabarmati riverfront. Authoritarian governance, neoliberalism, and water in Ahmedabad. Partecip Conflit 11(3):692–714
Pessina G (2022a) Environmental violence. In: Pellizzoni L, Leonardi E, Asara V (eds) Handbook of critical environmental politics. Edward Elgar, Cheltenham, pp 347–361
Pessina G (2022b) Fragility, environmental risks, and territorial safeguard. Transdisciplinary perspectives from inner areas. In: Italian National Network of Young Researchers for Inner Areas Committee (eds) Inner areas in Italy. A testbed for analysing, managing and designing marginal territories. ListLab, Trento-Barcelona
Pessina G (2023) Accidental ethnographies of energy transition. A project to decarbonise Italy's power stations as recounted by an urban planning consultant. Etnogr Ric Qual 2:247–272
Pickerill J, Chatterton P (2006) Notes towards autonomous geographies: creation, resistance and self-management as survival tactics. Prog Hum Geogr 30(6):730–746
Pileri P (2019) L'urbanista nell'antropocene. Sette questioni da cui partire per cambiare. In: Ferran F, Mattogno C, Metta A (eds) Coltiviamo il nostro giardino. Osare nuovi paesaggi, prendersi cura, inselvatichire il mondo. Derive Approdi, Roma, pp 97–106
Puig de la Bellacasa M (2011) Matters of care in technoscience: assembling neglected things. Soc Stud Sci 41(1):85–106
Puig de la Bellacasa M (2022) Embracing breakdown. In: Papadopoulos D, Puig de la Bellacasa M, Myers N (eds) Reactivating elements. Chemistry, ecology, practice. Duke University Press, Durham, NC, pp 195–230
Quarantelli E (1999) The disaster recovery process: what we know and do not know from research. http://dspace.udel.edu/handle/19716/309
Quarantelli E (2005) Catastrophes are different from disasters: some implications for crisis planning and managing drawn from Katrina. Social Science Research Council: Items. Accessed 6 Dec 2021. https://items.ssrc.org/understanding-katrina/catastrophes-are-different-from-disasters-some-implications-for-crisis-planning-and-managing-drawn-from-katrina/
Rajabi M (2025) Climate change, resilience and cultural heritage. In-between international debates and practical encounters. In: Polimi Springer briefs in applied science and technology. Springer, Cham
Ravazzi S (2024) Improvisation and emergency management: policy responses to COVID-19 in Italy. Palgrave Macmillan, Cham
Remes JAC, Horowitz A (2021 [1980]) Critical disaster studies. University of Pennsylvania Press, Philadelphia
Rivera DZ (2022) Disaster colonialism: a commentary on disasters beyond singular events to structural violence. Int J Urban Reg Res 46:126–135
Rivera Cusicanqui S (1997) Mujeres y estructuras de poder en los Andes: De la Etnohistoria a la Política. Escarmenar Rev Boliviana Estud Cult 2:16–25
Romeo S (2019) L'Acciaio in Fumo. L'ILVA di Taranto dal 1945 a oggi. Donzelli, Roma
Salvalai G, Quagliarini E, Blanco Cadena JD, Bernardini G (2024) Slow onset disasters. Linking urban built environment and user-oriented strategies to assess and mitigate multiple risks. In: Polimi Springer briefs in applied science and technology. Springer, Cham
Sistema Nazionale per la Protezione dell'Ambiente (SNPA) (2023) Rapporto "Consumo di suolo, dinamiche territoriali e servizi ecosistemici". Report 37. Available at: Consumo di suolo, dinamiche territoriali e servizi ecosistemici. Edizione 2023—SNPA—Sistema Nazionale Protezione Ambiente

Smith N (2006) There's no such thing as a natural disaster. Insights from the Social Sciences, 11 June 2006. Available at: https://items.ssrc.org/understanding-katrina/theres-no-such-thing-as-a-natural-disaster/

Solnit R (2009) A paradise built in hell: the extraordinary communities that arise in disasters. Penguin Books, New York

Staupe-Delgado R, Kruke BI (2018) Preparedness: unpacking and clarifying the concept. J Contingencies Crisis Manag 26:212–224

Sultana F (2024) Confronting climate coloniality. Decolonizing pathways for climate justice. Routledge, London, New York

Swyngedouw E (2015) Liquid power: contested hydro-modernities in twentieth-century Spain. MIT Press, Boston

Swyngedouw E, Heynen N (2003) Urban political ecology, justice and the politics of scale. Antipode 35(5):898–918

Vale L, Campanella T (2005) The cities rise again. In: Vale L, Campanella T (eds) The resilient city: how modern cities recover from disaster. Oxford University Press, New York, pp 3–26

Valent G (2019) Disaster recovery and violence in neoliberal times: community and spatial fragmentation in L'Aquila. Disaster Prev Manag Int J 28(4):446–459

Watts M (2015) Securing oil: frontiers, risk, and spaces of accumulated insecurity. In: Hannah A, Mason A, Watts M (eds) Subterranean estates: life worlds of oil and gas. Cornell University Press, Ithaca, NY, pp 211–236

Wisner B (2001) Risk and the neoliberal state: why post-mitch lessons didn't reduce El Salvador's earthquake losses. Disasters 25(3):251–268

Yin RK (2009) Case study research: design and methods. Sage, London

Zaccaria AM (2023) La Faglia Nascosta. Territori, Decisioni e Pratiche nel Cratere Sismico del 1980. FrancoAngeli, Milano

Zierler D (2011) The invention of ecocide: Agent Orange, Vietnam, and the scientists who changed the way we think about the environment. University of Georgia Press, Athens, GA

Chapter 2
Framing Extreme Events in Contemporary Territories

Abstract The chapter provides an extensive literature review on contemporary extreme events and emergencies with territorial relevance. The chapter explores how traditional disaster studies approaches have been challenged by emerging paradigms to understand extreme events. It identifies mainstream approaches in response to disasters and highlights their shortcomings. During the mid-twentieth century, the hazard paradigm dominated disaster research by treating disasters as events originating from natural forces outside human control, and proposing scientific technological advancements, zoning management, and forecasting as solutions. Based on critical social sciences trajectories—critical disaster studies, environmental justice, political ecology, urban political ecology and decolonial approaches—and delving deep into several contexts, the chapter frames disasters as socio-spatial products that cannot be understood if separated from existing inequalities and power unbalances. The case of Terra dei Fuochi in Southern Italy is instrumental in illuminating the nexus between extreme events and pollution. This chapter aims at bridging the gap between extreme events and critical social sciences scholarship that engages with social and spatial inequality, not limited to assessing impacts or offering recommendations.

Keywords Critical disaster studies · Environmental justice · Political ecology · Urban political ecology · Disaster capitalism · State violence · Environmental racism · Decolonising disaster

2.1 Mainstream Perspectives and Emerging Paradigms in Unpacking Extreme Events

This chapter explores how traditional disaster studies approaches have been challenged by emerging paradigms to understand extreme events. Traditional disaster studies have focused on technical management and hazard-focused methods to identify physical disaster causes and apply technical solutions for risk reduction and prevention. These approaches, originating from engineering, emergency management and several policy and planning sectors, prioritize risk assessment with

S. Armondi and G. Pessina, *Critical Spatial Thinking and Extreme Events*, PoliMI SpringerBriefs, https://doi.org/10.1007/978-3-032-18875-5_2

early warning systems and consolidate infrastructure resilience (Alexander 2018; Cutter 2016). During the mid-twentieth century, the hazard paradigm dominated disaster research, treating disasters as events originating from natural forces outside human control, while proposing scientific and technological advancements, zoning management, and better forecasting as solutions. Subsequent critical social sciences approaches—critical disaster studies, environmental justice, political ecology, urban political ecology, and decolonial studies—peppered with fresh research findings and empirical insights, revealed that social, economic, and political conditions determine disaster interpretation, define recovery measures and shape impacts on communities pushed to the margins (Wisner et al. 1994). Rather than a single body of theories or backgrounds, all of them are instead several independent research tracks being conducted, sometimes in different combinations and syntheses, and referring to different communities of practice in the late twentieth century.

Despite recent advances in academic debate, many post-disaster politics and planning responses are based on two main models that open understanding of the relationship frameworks between the state and its citizens. The first is the top-down model. It represents state-led interventions focusing primarily on "safe" recovery instead of promoting community bottom-up resilience (Tierney 2021). This model is implemented with the militarisation of post-disaster intervention, with the classification of space in safe zones and red inaccessible ones that sometimes aggravate the displacement of the population in the name of "security".[1] The second is the post-disaster model focused on economic redevelopment, which is mainly imbued with neoliberal strategies, promotes disaster response privatisation, and worsens social inequalities (Klein 2007; Pelling and Dill 2009). These two leading conventional models, sometimes intertwined, have received criticism for disregarding spatial and environmental justice principles and power relationships, but overlooking grassroots knowledge, which has triggered demands for decolonial and justice-centred disaster governance methods (Gaillard 2019).

Recent research works, as presented in the following sections, engage with changing dynamics, politicization and egalitarian aspirations, promoting community-focused disaster risk reduction methods. They do not ignore affected populations' agency and value Indigenous and local knowledge systems (Kelman 2020). The intensification of extreme events and disaster risks from climate change makes it essential to employ interdisciplinary, context-sensitive frameworks to reframe mainstream management strategies and their outcomes (Pelling and Garschagen 2019), also to unveil that development and climate change are two sides of the same coin.

[1] For example, Valent (2019) studied the post-disaster public intervention in Italy, after the L'Aquila 2009 earthquake, specifically examining the long-term fragmentation of the city's physical and social fabric caused by the planned displacement and resettlement of thousands of families in 19 semi-permanent housing developments situated in peripheral, rural areas of the municipality.

2.2 Disasters as Socio-spatial Produced Processes: Critical Disaster Studies

Critical Disaster Studies operate at the intersection of multiple disciplines to scrutinize established disaster management frameworks by examining how social and political histories determine vulnerability to shocks and crises. This understanding rejects hazard-focused and technocratic solutions by showing how disasters result from socio-economic frameworks established by capitalist city development, colonial histories, and systemic discrimination. Unlike most approaches, critical disaster studies begin with the idea that extreme events are not merely objective facts but, rather, are interpretive fictions, they shape the way people see the world (Table 2.1). Critical disaster studies uncover what is at stake by questioning the taken-for-granted concept of extreme event and arguing the standard categories that define people or places as vulnerable, resilient, or at risk.

Hewitt's works (1983, 1995) led disaster research to move from deterministic environmental explanations. Through the prominent book "*Interpretations of Calamity*," the geographer extensively criticises the dominant hazard paradigm, which prioritises physical factors over social and economic roots of disaster vulnerability. Hewitt's academic contributions demonstrate that disasters stem from long-term historical and structural processes that create unequal levels of vulnerability and recovery potential, rather than random events with no socio-political significance. With his analysis, Hewitt (1997) argued that mainstream disaster risk reduction models and post-disaster procedures obscure the primary causes of vulnerability by prioritising technical and managerial solutions over systemic transformation.

Through his foundational research in disaster sociology, Enrico Quarantelli, another cutting-edge scholar, disputed conventional views of disasters. In his expanded research, Quarantelli (1998) systematically challenged the misconception that extreme events always result in chaos by proving that communities demonstrate substantial self-organisation and mutual assistance after catastrophic events.

Table 2.1 How critical disaster studies challenge mainstream approaches

Mainstream disaster studies	Critical disaster studies
• Focused on hazards as natural	• Focused on social vulnerability
• Disasters as external shocks	• Disasters as historically shaped (colonial legacy and knowledge making)
• Emphasis on top-down governance	• Support to community-led recovery
• Relied on militarized responses (e.g., FEMA, Red Cross; Italian Protezione Civile)	• Critique of state violence and to support mutual aid
• Recovery framed as economic rebuilding	• Recovery conceptualised as territorial healing and spatial justice
• Focus on sustainability and resilience as personal responsibilities	• Critique of individual sustainability and resilience as a neoliberal concept

Quarantelli found that local community-based disaster preparedness systems function better than militarized and authoritarian approaches to building resilience. Critical Disaster Studies offer vital improvements to conventional research approaches through qualitative analysis that explores the relationships between power structures and socio-spatial inequalities concerning policy and planning activities.

Disaster is an "evolving concept" (Perry 2018); as a consequence, researchers must combine political economy analysis with historical materialism and critical frameworks to unpack disaster discourse from technical-managerial interpretations into understanding disasters as systemic social and environmental injustices. The book *What Is a Disaster? New Answers to Old Questions* (Perry and Quarantelli 2005) has been a pivotal text in critical disaster studies. Perry, Quarantelli and contributors question the established definition of "disaster" to criticise assumptions that view catastrophes solely as unconventional events that require technical or natural-scientific understanding. The authors focus on how disasters are socially constructed phenomena. They examine the power dynamics and structural institutions that influence vulnerability and recovery processes, specifically for vulnerable groups. Their pivotal research dismissed the idea of disasters as mere disruptions to everyday life by considering them as outcomes influenced by wider social and economic systems.

The text emphasises a critical research gap: to study the relationship between political entities and communities, it is crucial to consider their physical surroundings to reveal how disparities magnify the origins and effects of disruptive occurrences. By offering an interdisciplinary framework incorporating sociology, anthropology, and political economy, Perry and Quarantelli's edited book is inspiring to challenge how disasters are represented and framed, who defines them, the different mindsets involved, and whose voices are selected in policymaking. As a result, the volume remains a vanguard reference for critical disaster studies, encouraging further inquiry. Nonetheless, in Quarantelli et al. (2017), multiple case studies are employed to demonstrate how disasters affect areas differently based on local contexts, existing social power structures, and cultural standards.

2.3 Nature, Resilience and Sustainability: Contested Words in the Spatiality of Disasters

Critical geographers, such as Massey (1994) and Smith (1984), have accentuated the idea that space is socially produced and contested, rather than a neutral container within which disasters unfold. Neil Smith further built on this critique with his famous statement: "There is no such thing as a natural disaster" (2006). Smith concluded that natural disasters do not exist, because their impacts are always shaped by economic and political factors: pre-existing social inequalities forge the implications of environmental events like earthquakes, hurricanes, and floods. Furthermore, wealthy populations strengthen their resilience with better infrastructure, insurance options, and government assistance, but fragile groups remain unprotected.

According to Smith (1984, 2006), the concept of "the production of nature" provides a valuable lens to understand disasters from a unique viewpoint. Grounded in Marx's theoretical framework, Neil Smith initially advanced the "production of nature" thesis to examine the evolving interrelations between humans and their environments. Indeed, Smith contended that nature is not an a priori entity but rather is actively produced through socio-economic processes. By foregrounding this argument, Smith critically confronted—and thereby flustered—longstanding assumptions within both Marxist and non-Marxist approaches to nature. His view gained traction in a broader scholarly attempt to contest essentialist conceptualizations of nature and society, meant to "historicise socio-natures" (Loftus 2017). Smith argued that human activities, economic systems, and political decisions actively shape, use and abuse nature, rather than merely reacting to it. For instance, deforestation, land-use changes, and urban expansion driven by capitalist development have established conditions that aggravate the frequency and severity of floods, landslides, and wildfires. These transformations are not accidental but tied to powerful private economic interests and power structures, prioritising profit over "collective" environmental sustainability (March and Swyngedouw 2022).

Equally, the notion of "accumulation by dispossession," as deployed by the Marxist political geographer Harvey (2003), further illustrates how disasters often serve as an opportunity for spatial restructuring, thereby ensuring capitalist expansion. In many cases, post-disaster reconstruction efforts become opportunities for land grabs, gentrification, and privatization, and sometimes, criminal business, displacing vulnerable populations under the guise of resilience-building or safety.

Following the 2010 Haiti earthquake, for example, disaster recovery efforts led to the displacement of thousands of poor Haitians while elite-driven development projects flourished in newly cleared spaces (Katz 2013). Several critical studies in geography and political ecology have stressed how risks are distributed unevenly across all scales, from the bodies to regions, and to the planet. Spatial injustice generates vulnerability by forcing vulnerable and lower-class people into dangerous places with inadequate protection and resources. For instance, Indigenous areas experiencing violent resource extraction represent high-risk territories demonstrating this pattern. The systematic targeting of disadvantaged communities by environmental damage, through uneven disaster responses in planning and policy, results in repeated socioeconomic exclusion and unending poverty cycles. In this vein, the concept of resilience, commonly viewed as an objective or beneficial goal, has been defined by researchers "as a scientific notion operating behind an ideological disguise and [that] now faces critical examination" (March and Swyngedouw 2022, p. 3).

Several scholars warn that deploying resilience frameworks in disaster management policies may depoliticise risk reduction by placing the responsibility on affected communities instead of targeting the fundamental structural origins of vulnerability (Davoudi et al. 2012). The spatial dimension of resilience highlights how adaptive capacity is unevenly distributed, with wealthier regions and social groups often having greater access to protective infrastructure, insurance, and emergency response mechanisms.

In addition, to explain the connections between capital expansion and environmental injustice, other scholars have recently renewed Harvey's definition with the notion of "accumulation by contamination" (D'Alisa and DeMaria 2024), the process by which capital socialises the costs of contamination. Waste management has tremendous market potential and "emerges as a strategic realm for capitalists seeking to expand the scale and scope of accumulation" (D'Alisa and DeMaria 2024, p. 4) by all means (see the section on *Terra dei Fuochi*).

2.4 Environmental Justice and Racialised Disaster Vulnerability

Research in the Environmental Justice field consistently shows that disaster risks and recovery efforts reflect strong racial biases. The pioneer of environmental justice research, Bullard (1990), a scholarly activist based in Houston, bridged the gap between environmental issues and civil rights, discussing how disaster risks are affected by racial and class factors. The research conducted between 1980 and 1990 by Bullard demonstrated that African American neighbourhoods have been routinely targeted for the siting of toxic facilities in the cities of the South of the US. The research focused on the most critical social and economic impacts of the siting of noxious plants—municipal landfills, chemical plants, rubbish dumps, lead smelters—on Black communities. The author wrote that, since Houston did not have a zoning plan, he played the role of "researcher as detective" (Bullard 1990, p. 7). With interviews, on-site visits, and handwritten transcriptions of several documents, this research provided the basis for the research methodology of the Environmental Justice scholarship.

Furthermore, Pulido (2010, 2016), a geographer and scholar of Chicanx Studies, connected environmental justice issues and disasters to capitalism, indicating the central role of the latter in perpetuating historical patterns of environmental racism and exacerbating the environmental racism gap. Because "there is compelling evidence that environmental disparities between white and nonwhite communities (…) have not diminished" (Pulido 2016, p. 1), the author claimed both Environmental Justice activism and research have failed to correctly conceptualise "environmental racism as a constituent element of racial capitalism" (Pulido 2016, p. 1). Pulido explains that disasters function within a more extensive system where vulnerable, often nonwhite, communities experience persistently the worst effects of pollution and environmental damage. In contrast, affluent white communities receive protective infrastructure and economic aid. Pulido proposed confronting the state, rather than considering it a neutral force. The author noted that the State did not take substantial action to intervene in the environmental racism divide. Furthermore, the State is generally extensively committed to not solving the problem of environmental racism because it would be too expensive and damaging to industry, economy, and the balance of the political system and the State itself (Pulido 2016). A deep understanding of the basis of the inequalities and the devaluation of nonwhite bodies shapes the roots for changing disaster policy and response strategies.

The Difference Between Environmental Justice and Political Ecology Approaches

Political Ecology is another central research stream vital to understanding extreme events.[2] First, it is important to linger on the definition of this field of research. If there is a "political" ecology, this implies there must also be an "apolitical" ecology. Starting from two empirical examples—"When the bottom drops out of the coffee market, as it did in the late summer of 2001, what happens to the peasants who depend upon it and the forests in which it is harvested? When the World Bank helps to fund massive afforestation programs around the world, aimed at preserving tree cover and animal biodiversity, what actually happens to the hill forests designated for enclosure and the tribal people who live there?" (Robbins 2012, p. 25)—Paul Robbins highlighted the discrepancy between a political and an apolitical ecology, namely a difference between "identifying broader systems rather than blaming proximate and local forces; between viewing ecological systems as power-laden rather than politically inert; and between taking an explicitly normative approach rather than one that claims the objectivity of disinterest" (Robbins 2012, p. 25).

Even if, unlike Environmental Justice, Political Ecology has given industrial sectors comparatively limited consideration, Barca and Bridge (2015) considered industrialization as a crucial form of "slow violence" (Nixon 2011) that both reproduces and transforms inequalities into economic and political power, for which political ecology—which traditionally applies political ecological concerns to rural contexts—should undertake a new engagement.

Political Ecology research considers disasters through a critical framework, investigating how power dynamics and economic history shape environmental vulnerability, interpretation, and recovery processes. Although mainstream disaster studies treat disasters as separate natural occurrences, political ecology research examines how political and economic systems create disaster risks and differential effects. This scholarship provided novel insights into mainstream economic development theories whilst exploring their environmental connections and local conflicts, as Peet and Watts (1996) recognized.

Political Ecology studies illustrate, particularly in Global South territories, that extreme events emerge as socially constructed phenomena shaped by historical colonialism and capitalism and compounded by policy failures to direct which populations suffer the most. Political ecologists challenge neoliberal recovery models because they focus on economic expansion and elite advantages, rather than

[2] The definition of political ecology provided years ago by Blaikie and Brookfield (1987, p. 17) is arguably the most effective operational framework: "The phrase 'political ecology' combines the concerns of ecology and a broadly defined political economy." The first term, ecology, refers to a systems perspective on the relationships between land, soil, vegetation, and other organisms. The second term—political economy—is a broad concept, implying a Marxist, materialist view of capitalism, markets, social structures, and power.

community-driven resilience and environmental justice methods (Neumann 2005). Studies of hurricane recovery projects in Latin America and Southeast Asia exhibit how reconstruction efforts following disasters have aggravated land dispossession and increased socio-economic disparities (Oliver-Smith 2009). Political Ecology studies examine disasters by combining analysis of power struggles over land property and the privatization and commodification of resources to unveil the systemic conditions that worsen their impact on marginalised and lower-class groups.

According to Westcoat (2015), hazards research is a main rather than an ancillary topic in political ecology: "early political ecologists criticized hazards research for its behavioral emphasis on risk perception and decision-making to the neglect of structural power relations and differential vulnerability of social groups" (Westcoat 2015, p. 293). The Environmental Justice and Political Ecology approaches investigate the intersection of environmental issues and social inequalities. They differ in focus, theoretical foundations, scales, and methodologies. However, Environmental Justice and Political Ecology are both focused on environmental inequalities. Still, they use different scales and are contemplating diverse solutions: Environmental Justice centres on legal action and activist answers, and Political Ecology examines theoretical power dynamics along with historical and global environmental changes (see Table 2.2).

2.5 Urban Political Ecology: Rethinking the Production of Extreme Events

The connection between Urban Political Ecology (UPE) literature and extreme events is crucial to examine the role of cities in socio-environmental disasters and crises. The field of Urban Political Ecology is a relatively new research stream that emerged in the mid-1990s with the work of Erik Swyngedouw and Matthew Gandy, building upon earlier work in political ecology and urban studies. Urban Political Ecology research highlights how the sites of urban infrastructures, combined with state and urban policies, and capitalist growth, lead to unequal environmental risks throughout different parts of cities. It investigates the management of critical infrastructures as essential urbanization processes, i.e. water circulation, energy flow, capital allocation and labor activities. Swyngedouw (2004) and Heynen (2014) argue that extreme events function as integrated parts of socio-natural urban metabolic processes, rather than mere external disruptions. Researchers using urban political ecology analysis maintain that natural phenomena, such as floods, heatwaves, and hurricanes, do not exist in isolation as natural disasters. UPE scholars identify these events as socio-environmental processes that function through political, economic, and spatial systems. The pioneering studies conducted by Erik Swyngedouw and Maria Kaika reveal how political and "class" factors combined with capital investments influenced infrastructure location that produced unequal risk distribution. For instance, the perspective developed in groundbreaking books like *In the Nature of Cities*, edited

Table 2.2 Environmental justice and political ecology: some specific matters

Dimension	Environmental justice (EJ)	Political ecology (PE)
Main focus	Unequal distribution of environmental harms/benefits based on race, class, etc.	Political and economic structures that shape environmental degradation and social inequality
Primary goal	Legal and policy reforms to address environmental racism and ensure equity	Understanding and challenging systemic power structures and historical causes of environmental injustice
Scale of analysis	Local, urban and regional injustices (community-level)	Global, historical, and systemic (multi-scalar) analysis
Theoretical foundations	Civil rights movement, environmental activism, critical race theory, social justice, and environmental sociology	Marxism, post-structuralism, historical materialism, postcolonial theory, Foucauldian analysis, and political geography
Methodology	Participatory action research, case studies, GIS mapping, health impact assessments, and quantitative data collection	Historical and political-economic analysis, ethnography, discourse analysis, and critical cartography
Features of activism	Grassroots activism, legal action, community-based mobilization	Academic critique, systemic and structural theory development
Examples	• Toxic waste facility locations in Houston: documented racial disparities with African American communities targeted for the LULUs (locally unwanted land uses) • Flint water crisis: community activism and national media attention over contaminated water in a poor, predominantly Black city • Hurricane Katrina: highlighted racial and economic inequalities in disaster response and exposure in New Orleans	• Land grabs in Africa: studies on foreign agribusiness displacing rural farmers in Ethiopia and Mozambique • Neoliberal "preservation" in Africa: studies on the displacement of Indigenous communities for tourism and wildlife parks in Kenya and Tanzania • Lithium and copper extraction in South America: analysis of green energy transition fuelling new extractive frontiers and Indigenous dispossession
Alternatives/actions proposed	Legal action, environmental regulations, equitable urban planning, and civil rights lawsuits	Reframing environmental issues as systemic and political, advocating for economic justice, land reform, and alternative environmental policies

by Heynen et al. (2006), demonstrates how uneven historical development, alongside infrastructure neglect, leads to patterns of urban vulnerability.

Urban Political Ecology research provides essential analytical tools for exploring the economic and political portrayal of environmental extreme events. The approach situates extreme events within and beyond environmental justice "in the city" by following the power networks at play, historical political contexts, and infrastructure systems at different scales, thereby deepening social, political, and spatial dynamics. By challenging dominant narratives that portray disasters as apolitical external shocks, Urban Political Ecology studies, within a Marxist theoretical framework, disclose how urban infrastructures and urbanisation processes, alongside neoliberal governance and socio-environmental inequalities, produce vulnerabilities and risks and shape resilience.

Early Urban Political Ecology case studies, primarily in Global North contexts, examined technical approaches to disaster management, detailing the roles of local and State powers, as well as capitalist systems, in shaping policy and planning actions that influence urban vulnerability.

The field of Urban Political Ecology integrates Political Ecology principles within urban contexts to illustrate how cities and urbanization processes operate as dynamic arenas for both generating environmental risks and confronting these risks, rather than being mere sites of disaster occurrences (Heynen et al. 2006).

Investigations by UPE unveil how and why urban disasters affect populations differently, exposing regular trends of environmental injustice. On one hand, political and economic factors, particularly the operations of capital and State initiatives, create urban spaces that advantage select groups; on the other hand, they expose others to more significant environmental threats. Many urban low-income communities face environmental risks because they often reside in flood-prone territories, near industrial hazards or settlements that lack infrastructure. This condition stems from historical uneven development patterns, land speculation activities, and discriminatory spatial planning strategies. Political ecology and critical geography share the attempt to unpack how political and economic factors determine the development of socio-spatial disasters. These approaches reject the notion of disasters as extraordinary events by stressing the historical, political, and economic factors that define risk, vulnerability, and recovery. They aim to develop disaster methods that target the root causes of vulnerability, achieving fairness and justice rather than merely treating symptoms.

Furthermore, in their edited book, *Urban Political Ecology in the Anthropo-Obscene*, Swyngedouw and Ernstson (2019) led research towards addressing the—neglected—political dimension of global climate change, on one hand, through detailed inquiries of slow violence and contention of urban resilience; on the other, examining emergency governance practices in the questioned Anthropocene era.

Environmental Justice and Urban Political Ecology: Complementary but Different

Environmental Justice and Urban Political Ecology both examine power imbalances alongside spatial and environmental inequality. Environmental Justice underlines the role of local activism and legal solutions, unlike Urban Political Ecology, which adopts a structural theoretical foundation, detecting path dependency in historical long-term processes. Environmental Justice works primarily to address inequities through urban policies and activism. Urban Political Ecology primarily examines how urban economic and political forces contribute to environmental disasters on various spatial scales. Table 2.3 presents essential discussions concerning Environmental Justice and Urban Political Ecology.

Extreme Events and the Politics of Infrastructure

In general, infrastructures—including oil rigs and electrical wires, roads and water pipes, bridges, and critical infrastructure—"promise" circulation, distribution, and improved urban conditions (Graham and Marvin 2001). However, they are also affected by precarious assemblages and networks that threaten to break down and fail. Communities worldwide face ongoing problems of service delivery, ruination, and abandonment, and they employ infrastructure as a site both to make and contest political claims (Anand et al. 2018). The following sections briefly point out how

Table 2.3 Environmental justice and urban political ecology: key arguments

Dimension	Environmental justice (EJ)	Urban political ecology (UPE)
Core concern	Equity and justice in environmental issues. Unequal distribution of pollution, environmental burdens and benefits, especially among marginalised ethnic/Black communities	The "urbanization of nature" as a socio-material and hybrid metabolic system, to understand how social constructs like capital, money, norms, gender, and race become deeply connected with physical structures through built environments and urban green spaces, as well as water and other material flows such as food and energy
Focus	Environmental racism, pollution, climate justice, access to clean water, air, and green spaces	Spatial fix, political and economic powers shaping urban/nature issues, anthropogenic ecological change
Theoretical foundation	Rooted in civil rights activism, critical race theory, and social justice frameworks	Informed by Marxism, critical geography, and poststructuralist political ecology
Scale of analysis	Local to regional, often focused on community struggles and activism at the neighbourhood level	Uneven urbanization of nature, socio-ecological inequalities that pattern cities, and capitalist urbanization produced within, between, and beyond cities
Methods	Participatory action research, community organization, policy advocacy	Environmental history analysis, ethnography, discourse analysis of policy and planning, and case studies
Examples	Water crisis, air pollution in low-income communities, and urban heat island effects affecting marginalised groups	How privatization affects land access and urban water distribution, the commodification of nature, and infrastructure-driven inequalities

UPE research stresses the infrastructure's political role in shaping environmental risk and urban inequality dynamics.

Floods

In the last few years, Europe's territories have been severely affected by flooding, but the effects of flooding are distributed unevenly across regions. The serious May 2023 floods in Emilia-Romagna (Italy) that led to 23 rivers overflowing, devastating entire municipalities and causing over €8.5 billion in damages demonstrate how extreme weather events interact with both urban infrastructure weaknesses, soil sealing for industrial warehouses and logistics sprawl, and policy and planning shortcomings. The floods were presented as "natural" disasters. However, they resulted from prolonged urbanization sprawl, land take actions, soil sealing into flood-prone zones, inadequate drainage investments, and reduced regulatory control, marking modern neoliberal urban management practices. Economic and political processes influence the impact of floods and the distribution of risk.

The trends observed in Slovenia's 2023 floods affected two-thirds of the country and forced the displacement of approximately 8000 residents. Severe flooding affected Norway and Sweden that year, while a hydroelectric plant in Norway partially collapsed, exacerbating the situation. The incident demonstrates that inadequate energy infrastructure regulation makes it a significant disaster threat. Urban Political Ecology literature, as demonstrated in several case studies, indicates that infrastructure systems operate within socio-political frameworks that distinguish between those who receive protection and those who face exposure (Kaika 2004). Furthermore, fragile communities suffer the most during flooding due to ineffective evacuation strategies and delayed emergency responses (Swyngedouw 2004).

Droughts and the Uneven Politics of Water Access

Droughts are reshaping urban and rural geographies across Southern Europe, with significant consequences for agriculture, industry, and public health. In May 2024, Italy, Spain, and Greece faced one of the worst drought periods in recent history, with 70% of Greek territory under a drought alert and Italy's Po River basin experiencing dire water shortages. The effects of droughts are not merely about the lack of water; they are about who controls and distributes it.

Economic interests have often been prioritised over the needs of vulnerable groups in drought management. In Spain, for example, large agribusiness firms and urban tourism centers continue to receive priority access to water. In contrast, small farmers and rural communities face severe restrictions. This reflects what Swyngedouw (2015) labelled "hydro-social relations", where water scarcity is not just about physical availability but is affected by different powers at stake, attempts of privatization, and operations of capital that can affect access and availability.

Water management has disproportionately affected the smaller municipalities and working-class or subaltern groups in several countries. At the same time, wealthy districts and corporate agricultural operations often negotiate private water access agreements, reinforcing socio-spatial inequalities. These droughts uncover the multifaceted contradictions of multi-scalar water governance, where climate adaptation policies usually reinforce market-based allocation rather than prioritising social equity.

"Political" Wildfires. Climate Crisis and Planning Approaches to Wildfire Risks

The 2023 heatwave, known as "Cerberus," caused temperatures of 47 °C in Sardinia and 41.8 °C in Rome, heightening wildfire risks and straining public health systems. Rising temperatures, alongside prolonged droughts and changes in land use, have increased wildfires across Southern Europe. Spain, Greece, Italy, and Portugal suffered the burning of over 500,000 ha of land in 2023, resulting in widespread displacement, declining air quality, and loss of biodiversity. Climate change and neoliberal land policies have made wildfires a significant political issue.

Urban Political Ecology studies argue that wildfires are disasters, sometimes arson, but they are also largely connected to the turbulent expansion of urban development, poor land management practices, and extensive deforestation. Wildfire fuel loads have increased in Italy and Spain due to rural depopulation and speculative urban development, which have reduced traditional land management techniques that previously mitigated fire hazards, such as controlled burns and agroforestry.

According to Keenan et al. (2018), the notion of "climate gentrification" represents a key to understand the link between real estate speculation and the climate crisis. The increasing threat of wildfires in peripheral spaces around Barcelona, Athens, and Rome (in 2024 and 2025) has led to zoning restrictions that forced out low-income residents, while simultaneously promoting redevelopment projects led by the elite.

In "conventional" Global South cities, marginalized and lower-class urban populations face disproportionate harm from heatwaves and wildfires because they live in inadequately insulated housing and informal settlements or lack green space. Sometimes green infrastructure interventions produce or exacerbate urban socio-spatial inequities in self-built settlements (Anguelovski et al. 2016). Environmental inequalities grow more substantial when the necessary resilience tools for disaster response become exclusive commodities accessible solely to privileged groups. Both in the Global South and Mediterranean climate zones, the local knowledges of guarding and protecting forests, pastures and fallows through human presence are often in contrast to State narratives of combating and suppressing wildfires (Pyne 2009).

The increasing occurrence of floods, droughts, and wildfires throughout European cities demonstrates the profound structural disparities between climate adaptation and disaster management systems. Disasters occur from uneven urbanization processes alongside infrastructure investment decisions and territorial exploitation under neoliberal governance, instead of being natural events. Paradoxically, on one side, acute environmental disasters, such as floods and wildfires, often receive public attention; on the other side, slow-moving environmental problems, including toxic pollution and waste mismanagement, pose significant challenges to social justice and climate adaptation.

2.6 Decolonizing Extreme Events: Beyond Conventional Post-disaster Recovery

The theoretical framework of extreme events often incorporates Eurocentric and US technocratic perspectives, which uphold colonial histories, sidelining Indigenous knowledge systems and promoting capitalist reconstruction methods. Decolonising disaster studies requires examining historical and political contexts alongside economic structures and knowledge-production processes that shape vulnerability and recovery systems; it involves investigating how colonial legacies and neo-colonial development approaches define disaster governance; finally, it calls for studying Indigenous sovereignty-based strategies, environmental justice, and community-driven resilience methods to challenge these systems.

Colonial histories have shaped mainstream disaster management frameworks, which determine who holds power over knowledge creation and decision-making processes, as well as the distribution of resources. Extractivist economic systems, combined with forced evictions, displacements, relocations, and externally driven infrastructural developments, continue to dominate disaster-prone areas within the Global South, disregarding local needs. International aid agencies, multinational corporations, and state actors collectively frame top-down recovery responses for post-disaster operations, which risk replicating colonial practices. They are mainly affected by an epistemic injustice gap. Indigenous knowledge practices to enhance disaster preparedness and recovery efforts apply vernacular methods tailored to specific local contexts. However, mainstream disaster management systems disregard Indigenous knowledge paradigms and compel local communities to depend more on Western scientific methods.

Decolonization requires recognition of Indigenous knowledge, aimed at promoting land-based approaches to maintain a balance between society and nature. According to decolonial research, local groups are crucial for combining traditional Indigenous understandings with grassroots measures to redefine resilience and take charge of new research and disaster risk management.[3]

Extractivism[4] is a central notion for examining contemporary extreme events approaches with a decolonial lens. There is a growing recognition of the limitations of reconstruction initiatives led by partnerships between global corporations and State governments, as they tend to prioritise economic gains over environmental and social justice principles. For example, the intervention strategy established after L'Aquila

[3] Linda Tuhiwai Smith defined the term "research" as "one of the dirtiest words in the Indigenous world's vocabulary" (2021, p. 1), mainly because it has long been associated with Western empire-building and colonialism.

[4] Despite the ubiquity and opacity of the term that often ends up identifying every process, everywhere, through which value is generated for capitalism, in Latin American studies, where the term has been coined, extractivism again "names a mode of economic production in which natural resources are taken out of one geographic location – often on a colonial periphery – and utilised as raw materials, that is, as inputs to industrial processes elsewhere" (Szeman and Wenzel 2021, p. 506).

earthquake in 2009 in Italy, can be condensed in a combination of deregulation policies and the parallel application of strict control: "A dual-action system, typically neoliberal, was deployed, consisting of a policy of 'non-interference' towards business combined with heavy state interventions aimed at preventing or repressing social action" (Valent 2019, p. 447).

In this context, the work of sociologist Maristella Svampa, among others, has been central in identifying and portraying the conceptual features of extractivism. Seen as "a structural trait of capitalism as a world economy" (Svampa 2019, p. 15), the extractivism conceptualization has experienced significant changes and discussions in the twenty-first century, driven by several scholars, especially in Latin America. Furthermore, Svampa's theorization of "territorial emptying" (Svampa 2013) introduced a focus on the main geographical specificities. Drawing on Svampa, Caroca (2025, p. 3) indicates that it can be "defined as the deliberate disappearance (or erasure) of the socio-material features that compose a territory, in order to facilitate the extraction of natural resources and/or the production of surplus value". Introducing main geographical specificities in Svampa's conceptualization, first, there is a slighter accent on race and the colonial legacy and more emphasis on local socio-economic features, such as poverty, geographic isolation, social fragility, and land productivity (Svampa 2014, p. 150). Second, Svampa steps back from the opposition of settler/hegemonic and Indigenous/subaltern actors to take stock of the power of the three main actors: corporations, nation-states, and social organizations.

Svampa (2019) also highlights alternative concepts and theories that explore the cultures of Latin American socio-environmental movements, such as the Andean concept of *Buen Vivir* and *Derechos de la Naturaleza*, advocating for the end of the "development" and "underdevelopment" era, due to its ambiguities and limitations, and anticipating a new political cycle. The author argues that post-disaster reconstruction efforts often lead to the initiation of new extractivist activities, which exacerbate environmental conflicts and diminish local empowerment. As a result, humanitarian aid and post-disaster relief operations frequently become tools for neo-colonial objectives, sustaining dependencies and exerting geopolitical influence.

Duffield (2001), in a book focused on the changing nature of North–South relations, highlighted the incorporation of security and conflict resolution into the development discourse of aid policies. If "underdevelopment has become dangerous" (Duffield 2001, p. 4), then militaristic top-down approaches, complete demolition, development financing, and privatization models often target disaster-affected regions to influence them and further economic disparities and resource extraction. For example, Schuller (2016) analysed Haiti's earthquake, observing that recovery operations were led predominantly by foreign NGOs and international donors, which left local organizations while creating economic dependencies. Imperiale and Vanclay (2020) focused on reconstruction interventions following the 6 April 2009 earthquake in L'Aquila, where local political leaders assumed the role of civil protection authorities. The authors shed light on how local authorities also applied the command-and-control approach, aggravating local social and environmental risks and impacts, hindering local communities from learning and limiting their contributions to post-disaster interventions.

How can the decolonial option alter disaster aid policy and mainstream approaches to emergency response? A shift to a potential post-development conceptualisation,[5] accentuating alternative solidarity-based approaches that focus on local leadership, mutual aid networks, and equitable resource distribution, should replace patriarchal and colonial models in disaster responses. Extreme events result from historical socio-spatial inequities, environmental degradation, and capitalist exploitation; consequently, the reactions should involve or be driven by vulnerable groups who experience the impact directly, ensuring their input in the recovery process instead of implementing external solutions. According to Rob Nixon's concept of "slow violence", industrial pollution, new infrastructure construction (dams and plants), and deforestation, gradually build up and primarily affect and displace poor and ostracized populations (Nixon 2011). Policymakers can recognize bias and reduce structural inequalities within post-disaster policy by decolonising disaster studies and managerial responses. Equitable post-extreme events actions demand fresh theoretical methods and tangible modifications to policy-making processes, grassroots movement strategies, and aid distribution frameworks. Decolonized disaster responses enable communities to take control over their futures, reclaiming their lands through resistance to neoliberal regulations and neocolonial control. Academic research demonstrates the direct connections between resource extraction—from labour, bodies, and nature—and environmental destruction, both through contemporary capitalist global systems and colonial histories.

Researchers and activists have developed various theoretical frameworks explaining the intertwining and assemblage of processes that operate together. Gómez-Barris (2017) coined the term "extractive zones" to describe territories, whereas colonial economic profit and capitalist exploitation result in systematic environmental crises and ongoing eviction and displacement. The geographies of extraction represent intricate socio-ecological knots that emerge from the combined actions of governments and corporations that systematically exploit natural environments and human communities. Extractive zones usually exist in territories where Indigenous and racialized communities live, and environmental rules are poor, while resistance faces criminal charges and violent suppression. Gómez-Barris traced these zones that occur within colonial pasts and capitalist systems, both capable of exploiting specific landscapes and communities "as disposable".

However, Gómez-Barris highlighted that extractive zones represent more than devastation sites as they function as spaces where traces of possibilities and glimmers of resistance exist. Despite the destructive nature of extraction practices, alternative "ways of being, knowing, and relating to the environment manage to persist and emerge" (Macklin 2022). Extractive regions become platforms to nurture cultural renewal, political action, and knowledge-based challenges. Gómez-Barris explained

[5] According to Kothari et al. (2019), editors of the Post-Development Dictionary, the notion is related to three framework: (a) the possibility to question capitalism's ability to fully and naturally occupy economy; (b) the possibility to contest growth, tempering the role of economy and (c) the chance to challenge patriarchy, contrasting it as taken for granted primacy of masculinist approaches to political leadership and social privileges.

that Indigenous communities engage in practices that stand against extractivist principles. The Mapuche people in southern Chile are a notable example of resistance because they have consistently opposed forestry industries and extractive projects that threaten their ancestral territories. The Mapuche have maintained their ecological knowledge systems by defending their rights through land occupations and legal battles, as well as cultural revitalisation efforts, to oppose the mainstream development model.

These liberatory actions interrupt extractive processes by offering new ways of connecting with land, community, and more-than-human entities. Extractive zones operate as dual frontiers. On one side, they are spaces of accumulation that transform ecological life into market value, hiding the violent foundations of this conversion. On the other hand, they support decolonial futures that develop through relational ecologies and reciprocal care practices. Extractive zones hold a hauntingly transformative potential, enabling them to question prevailing systems and create new visions of existence that transcend capitalist and colonial structures.

The concept of "rendering land investible" by Li (2014) becomes the first key component in our approach, by showing how environmental crises and disasters shape the process of transforming land into a standardised investment asset. In her essay, *What is the Land?* As an anthropologist, Li investigated the tension of actively restructuring land within intricate social and ecological networks to make it attractive and valuable for global capital markets. Legal reforms and cadastral mapping work alongside titling and zoning to detach land from its historical users and ecological connections. The objective is to make the land appear as "empty," "underdeveloped", "underutilized," or "available", and this often happens after extreme events. Through this transformation process, land becomes a neutral asset that can be enclosed and extracted from which commodities can be produced.

Li demonstrated that reassembly operations move beyond technical neutrality by driving the shift of systemic risk and creating environmental instability. Making land a viable investment option promotes speculative land acquisitions, plantation-based economies, and extensive extraction operations, resulting in significant alterations to land use, including deforestation, biodiversity destruction, and displacement. These factors become fundamental to other socio-environmental crises. The altered landscapes experience increased exposure to gradual ecosystem declines and direct environmental catastrophes, such as floods and fires.

Li noted that Southeast Asia experienced significant changes in local hydrology, including damaging seasonal haze and soil erosion, following substantial agricultural investments that converted large areas into palm oil plantations. Land changes defended by development and productivity narratives emerged as central locations for environmental damage. "Investability" of land represents both a spatial and political undertaking that amplifies disaster vulnerability and veils responsibility using global finance and technocratic planning language.

It is important to note that in Li's theoretical suggestion, environmental disasters should not be regarded merely as adverse, incidental events to handle, but rather as outcomes of spatial changes driven not only by capital interests. Indeed, her research demonstrates that routine government procedures, such as title registration and investment incentives, are key to transforming global ecological systems. The relationship between disaster creation and land spatial politics, combined with investment in infrastructure, becomes much clearer through this understanding. Indeed, neoliberal capitalism has facilitated the transformation of ecosystems into value-producing sites as part of the commodification process of "cheap" nature (Moore 2015).

The second key component—the reverse of investability—is the concept of "wastelanding", extensively examined in the field of settler colonial studies. Traci Brynne Voyles describes wastelanding as "a fully colonial project of rendering resources extractable and lands and bodies pollutable" (Voyles 2015, p. 25). Consequently, entire "territories, bodies, worldviews, epistemologies, histories, and cultural practices are all equally rendered as waste" (Voyles 2015, p. 11). Voyles describes the causes of wastelanding as external to the material conditions of the land, suggesting instead the role of forms of representation and political rationality such as (perceived) beauty, or causes stemming from social, cultural and legal domains. Wastelanding is, primarily, an enactment of "environmental imagination that sees [certain lands] as threatening, marginal, and [...] 'barren' places" (Voyles 2015, p. 8).

Brenner and Katsikis (2020) named these spaces the "hinterlands of the Capitalocene". They illustrated how their seemingly "peripheral" position remains structurally crucial to supporting capitalist growth. Social movements and radical environmental activists have begun to use the term "sacrifice zones" (Juskus 2023) to describe places where communities endure ecological destruction alongside institutional neglect. Although separate in their methods, these two concepts and approaches both critically delve into how spatial and political factors create unequal environmental vulnerability in their research trajectories.

Environmental catastrophes and degradation occur unevenly across regions, and their causes extend beyond simple technical failures; they expose a deeply rooted system that foregrounds capital accumulation. To cope with contemporary extreme events, multiple disciplines must investigate the political-economic frameworks and historical developments that have produced a cascade of crises, overlooking long-term ecological sustainability and social justice issues. Scholars have focused on the climate crisis alongside related crises, revealing broader environmental injustice issues tied to the uneven patterns of capitalist modernity, colonialism, and the role of pollution. Emblematic is the book *Pollution is Colonialism* (2021). The author, Max Liboiron, highlighted that extraction logic and displacement methods persist, combined with active colonial exploitation and eviction. The author described how infrastructural and industrial operations, by exploiting land and water, reproduce colonial power structures. Colonial knowledge systems, combined with mainstream environmental policies, shape which knowledge forms hold value, thereby determining the lands that face exploitation. Consequently, environmental damage serves as a permanent consequence of colonial domination, rather than purely a result of inadequate management.

It is crucial to reassess disaster interventions through this critical perspective by shifting from technocratic approaches to structural transformation. This viewpoint prompts us to examine who bears the cost of economic benefits and how capitalism transforms disasters into an everyday reality. Liboiron's critique of environmental science's colonial roots supported movements to remove colonial influences from disaster discourse. The research demonstrated that prevailing environmental policies often marginalised Indigenous and "switch off" local knowledge systems, promoting scientific rationalism with foundations in settler colonial practices. Decolonial futures require both physical change and the overcoming of dominant knowledge systems that justify environmental destruction.

2.7 Defining Disasters Between Anthropocene and Capitalocene

The interplay between economic, legal and institutional structures, which advance the privatization of natural resources and deregulation, but shift environmental costs to external parties, maintains this dynamic. Harvey's (2003, 2005, 2007) renowned theory of "accumulation by dispossession" explained how environmental risks grow through privatization, extraction of resources and industry deregulation, valuing capital accumulation above ecological and social health. The relentless pursuit of urbanisation of capital, without regard for consequences, has exacerbated environmental crises that emerge mainly in Latin American mining operations, Southeast Asian agro-industry expansion, and Arctic fossil fuel extraction due to their integration into worldwide production networks and financial tools.

Moore (2015) introduced the concept of the Capitalocene as a provocative alternative to the Anthropocene's crudely universalist narrative, as the latter broadly attributes ecological problems to human actions rather than capitalism's distinct historical and structural development patterns, seemingly to be more—and rather loosely—invoked as a new synonym for modernity. According to Moore, neoliberal capitalism has enhanced nature's commodification process, turning ecosystems into places for generating surplus value: capitalism functions as a "world-ecology" which incorporates nature within its accumulation and exploitation operations, leading to environmental destruction becoming an essential component instead of an incidental result of economic development. The Capitalocene framework posits that capitalist world ecology is the primary driver of environmental problems, suggesting that capitalist development has altered planetary systems, leading to increasingly severe disasters. Malm (2016) and other eco-Marxist scholars demonstrate how fossil capital drives climate instability and economic downturns and claim that fossil fuel-dependent capitalism expansion fails to meet environmental sustainability criteria. The joint analysis of these theories reveals how systemic forces produced risks and catastrophes during the Capitalocene, rather than accidental events or human errors. Their existence fundamentally relies on the logic of capitalist production processes.

The Capitalocene is advocated as the proper term to describe current planetary urbanization by urban thinkers who study the relationship between capitalist systems and ecological crises. The notion rejects the Anthropocene's general human-caused environmental change narrative by showing that political and economic systems have driven ecological instability and disasters. The Capitalocene enhances critical studies in global political economy regarding environmental issues by identifying the multi-faceted factors that lead to unequal distribution of costs and benefits from how nature is managed and exploited within capitalist production and consumption systems and global value chain (Saurin 2001; Sovacool and Linnér 2016). The growing acceptance of the Capitalocene concept in the social sciences highlights the development of stronger connections between this framework and the field of environmental studies, with a focus on critical disaster and risk studies.

To conclude, research indicates that the expansion of global capitalism through industrial production and consumption systems has intensified human-induced environmental changes that characterise our modern era, including the COVID-19 pandemic (Kallis 2021; Malm 2020). Furthermore, the Marxist scholar Seymour (2024) explained that far-right movements exploited the environmental crisis as a tool for political gain by inventing the term "disaster nationalism". This is instrumental in describing how these groups present catastrophic solutions to social problems and merge ecological disasters with their political strategies. These perspectives demonstrate that global capitalism structures contain the foundational risks and disasters of the Capitalocene, which demand critical analysis of both our political and economic systems to tackle environmental emergencies effectively.

Disaster Capitalism and Critical Environmental Justice

The neoliberal governance framework actively shapes people's vulnerabilities and, consequently, their responses to disasters. According to Peck and Tickell (2002), due to the fact that neoliberalism transforms urban governance through market-based approaches, commodification and privatization, concurrently reducing or transforming state involvement in social and environmental safeguards, the impact of neoliberal governance extends deeply into disaster preparedness and recovery processes. The financial sector's connection to ecological harm through speculative fossil fuel extraction and agro-industrial investments worsens instability by creating a reinforcing cycle of economic collapse and environmental disaster. The prioritization of immediate profit by capitalist systems over sustainable development leads to increased worldwide disasters, creating an unpredictable and unstable future for all forms of life. Klein's (2007) disaster capitalism theory helps explain how disasters become opportunities for elite wealth accumulation and privatization tactics. Indeed, significant disasters enable governments and corporations to use emergency measures for to sell public assets, simultaneously stripping away social protections and forcing vulnerable populations out to make room for speculative urban redevelopment. New Orleans experienced extensive privatization of public housing and education following Hurricane Katrina (2005), which adversely impacted Black and low-income residents who could not return to their dwellings because of increased living expenses (Adams 2013).

2.8 Terra dei Fuochi: Toxic Pollution as Political, Colonial, and Ecological Disaster

Immediate public concern about acute environmental disasters, e.g. floods and wildfires, contrasts with long-term, silent threats, e.g. toxic pollution and waste mismanagement, which equally challenge environmental justice issues. Terra dei Fuochi in southern Italy is Europe's most emblematic example of environmental degradation intersecting with organised crime and governmental breakdown. By 1995, in the Campania region, between Naples and Caserta, the Terra dei Fuochi case had experienced waste trade between the Camorra (the regional criminal organization) and corporations across Italy and Europe, prolonged illegal hazardous waste dumping due to Camorra activities, weak regulatory systems, and industrial deregulation. The situation worsened public health emergencies, damaging local ecosystems and strengthening socio-spatial inequalities, which affected peripheral neighbourhoods and lower-class communities the hardest. Some scholars have defined the persistent environmental crisis in this territory as "environmental crime", exhibiting how waste is deeply connected to structural political and social injustices (Armiero and D'Alisa 2012; Armiero 2021). In the Terra dei Fuochi case, dumping and pollution function as a political and State spatial construct rather than a simple consequence of waste disposal activity. Due to its subaltern position, the region is a peripheral "sacrifice zone" within national and European political economies. At the same time, environmental degradation becomes normalised alongside, firstly, a denial. Denial can be traced back to Cohen definition as "assertions that something did not happen, did not exist, is not true or is not known" (Cohen 2001, p. 3). According to Pulido (2024), denial follows when the truth is too disturbing, frightening or expensive to be accepted. This scepticism, coupled with inconsistent regulatory intervention for decades, prevented any effort to address the environmental pollution. Secondly, with the devaluation of lower-class bodies, blame is placed on the behaviours of the inhabitants of Campania. The devaluation of lower-class bodies has been incorporated into economic processes that include pollution (Pulido 2016).

The occurrence of pollution-related diseases, including respiratory problems and cancer, highlights the systemic environmental injustice in the region, leading to an extreme event that has been experienced as a "victimful" crime, according to D'Alisa et al. (2017).

The widespread incineration of hazardous waste—*i fuochi*—has escalated air pollution levels and simultaneously increased local climate extremes that strengthen urban heat islands and cause disproportionate damage to vulnerable groups such as children and older people. Soil degradation diminished water retention abilities and boosted runoff, elevating flood dangers, and widespread water contamination made local sources unusable for farming or household needs, thus worsening drought vulnerabilities. Burning hazardous waste produces dioxins and additional pollutants, intensifying public health emergencies while speeding up regional climate change effects. Only with Decree 136/2013—later converted into Law 6/2014—the Italian state began, with extreme delay, to take a systematic interest in the issue. The provisions of that decree mandated mapping, contaminated farmland, provided resources for environmental monitoring and health screening and enhanced controls on the territory. Only in 2015, environmental crimes were finally codified in the criminal code with Law n. 68/2015, a legislative measure named *Ecoreati*. This made environmental violations illegal under Italian law, alongside enhanced checking systems and public activism movements like *Stop Biocidio Coalition*[6] (De Rosa 2017).

Greater public awareness and increased prosecutions resulted from these measures. However, enforcement shows inconsistency and land reclamation efforts continue to progress too slowly due to insufficient funding. Large sections of the region resulted in large-scale contamination, even with support from EU environmental remediation programs and local organic farming activities. Environmental monitoring advancements have established a connection between pollution exposure and cancer prevalence, yet institutional responses are delayed. In the meantime, healthcare access for affected communities remains inadequate.

The Terra dei Fuochi extreme event is emblematic. It illustrates that environmental disasters have both a political and ecological dimension, and they can develop through prolonged political neglect and systemic injustice, rather than abrupt shocks. The production of disasters results from social and political processes influenced by historical patterns of lacking governance alongside deregulation and organized crime. Policy responses focusing on technological surveillance and incomplete cleanups, ignoring socio-economic foundations, continue to fall short of effectiveness. On January 30, 2025, the European Court of Human Rights in Strasbourg issued a ruling that rewrites the official history of the Terra dei Fuochi,[7] because the Court condemned the Italian State for disregarding the issue. Two essential insights occur from this case:

1. Disasters are socially and politically produced in relation. Terra dei Fuochi represents a human-made disaster created by weak institutional systems, which span several decades and are exacerbated by both systematic "wastelanding" and "denial"[8] of the pollution by the Italian state, neoliberal deregulation policies and criminal organisations
2. Policy responses must address root causes. Technocratic measures, including surveillance and limited clean-ups, cannot solve environmental problems unless accompanied by fundamental changes in the economic and political systems. Technical solutions that failed in Terra dei Fuochi exemplify broader patterns in

[6] The role of the Campania committees of the first decade of the 2000s should be emphasised because they disputed against the disastrous urban waste plan and denounced illegal dumping. In the summer of 2012, the *Coordinamento Comitati Fuochi*, a network of more than fifty Campania committees against pollution, was organised (De Rosa 2025).

[7] The proceedings, which began in 2015, were triggered by the complaint of forty-one Campania citizens and five local associations against the Italian State for jeopardising their right to life.

[8] Pulido (2024, p. 59) in a chapter on wildfire and the role of denial of climate change in a polarised political setting, underscores that "Denial ensues when the truth is too disturbing or threatening to be accepted".

Italian State disaster management that point out infrastructure and technology, while overlooking social justice issues.[9]

The illustration of Terra dei Fuochi has some main takeaways. Research must move beyond limited technical solutions to fundamentally contest stagnant governance structures and merely economic considerations. Agroecology-based community land restoration initiatives effectively restore polluted territories and stand against speculative, criminal waste management and urban development, "whatever it takes". Effective environmental policy and planning tools should include more rigorous regulatory enforcement and improved clarity in industrial and waste management operations to achieve sustainability. The State must prioritize restoration actions, including full healthcare coverage for diseases caused by pollution, to address the sustained socio-environmental impacts of contamination.

2.9 Conclusion

With this chapter, we wanted to look anew at some of the assumptions and premises that have sometimes undergirded arguments about literature trajectories and extreme events. Studies in Environmental Justice, Political Ecology, Urban Political Ecology, critical geography, and decolonial perspectives demonstrate that extreme events represent intensified socio-political conflicts over space control, resource allocation, knowledge production and economic and political systems. These approaches challenge mainstream disaster approaches by illustrating how urbanization alongside infrastructure and neoliberal policies shape spatial disaster vulnerabilities rather than treating disasters as technical problems devoid of political context. More recent studies have, in some ways, blurred the separation between EJ, PE, and UPE, with the definition of a field of study that attempts to overcome some of the limitations of previous thinking, particularly in the context of Critical Environmental Justice studies (Pellow 2018). Finally, all these research trajectories advance essential backgrounds to craft inclusive urban futures by examining justice, equity and power alongside knowledge production in disaster research, planning and policy.

References

Adams V (2013) Markets of sorrow, labors of faith: New Orleans in the wake of Katrina. Duke University Press, Durham

Alexander D (2018) Natural disasters. Routledge, London and New York

---

[9] D'Alisa et al. (2010) focused on the complexity of the issue at stake in the Campania case conflict and concern about democracy. They asserted that the decision-making process adopted by subsequent Italian governments, alongside repressive laws, oversimplified a wicked problem and hidden emergent perspectives and values.

Anand N, Gupta A, Appel H (eds) (2018) The promise of infrastructure. Duke University Press, Durham
Anguelovski I, Irazábal-Zurita C, Connolly JJ (2016) Grabbed by the green: the changing nature of green gentrification. Urban Stud 53(8):1542–1560
Armiero M (2021) Wasteocene: the political ecology of waste and toxic landscapes. Cambridge University Press, Cambridge. https://doi.org/10.1017/9781108920322
Armiero M, D'Alisa G (2012) Rights of resistance: the garbage struggles for environmental justice in Campania, Italy. Capital Nat Social 23(4):52–68
Barca S, Bridge G (2015) Industrialization and environmental change. In: Perreault T, Bridge G, McCarthy J (eds) Routledge handbook of political ecology. Routledge, London and New York, pp 366–377
Blaikie P, Brookfield H (1987) Land degradation and society. Methuen and Co. Ltd., London and New York
Brenner N, Katsikis N (2020) Operational landscapes: hinterlands of the Capitalocene. Archit Des 90:22–31. https://doi.org/10.1002/ad.2521
Bullard R (1990) Dumping in Dixie: race, class, and environmental quality. Westview Press (2018, Routledge, London and New York)
Caroca A (2025) The monster has landed: shifting land tenure regimes and the political ecology of a Chilean mining 'wasteland'. Geoforum 159:104202
Cohen S (2001) States of denial: knowing about atrocities and suffering, Print, https://www.wiley.com/en-gb/States+of+Denial%3A+Knowing+about+Atrocities+and+Suffering-p-9780745623924
Cutter SL (2016) The landscape of disaster resilience indicators in the USA. Nat Hazards 80(2):741–758
D'Alisa G, Demaria F (2024) Accumulation by contamination: worldwide cost-shifting strategies of capital in waste management. World Dev 184
D'Alisa G, Burgalassi D, Healy H, Walter M (2010) Conflict in Campania: waste emergency or crisis of democracy. Ecol Econ 70(2). https://doi.org/10.1016/j.ecolecon.2010.06.021
D'Alisa G, Germani AR, Falcone PM, Morone P (2017) Political ecology of health in the land of fires: a hotspot of environmental crimes in the south of Italy. J Polit Ecol 24:59–86
Davoudi S, Shaw K, Haider LJ, Quinlan AE, Peterson G, Wilkinson C (2012) Resilience: a bridging concept or a dead end? Plan Theory Pract 13(2):299–333
De Rosa SP (2017) Reclaiming territory from below: grassroots environmentalism and waste conflicts in Campania, Italy. Dissertation, Lund University
De Rosa SP (2025) La sentenza sulla Terra dei Fuochi e l'archivio delle lotte ambientali, 4 Feb 2025. Napoli Monitor. https://napolimonitor.it/la-sentenza-sulla-terra-dei-fuochi-e-larchivio-delle-lotte-ambientali/
Duffield M (2001) Global governance and the new wars: the merging of development and security. Zed Books, New York
Gaillard JC (2019) Disaster studies inside out. Disasters 43(S1):S7–S17
Gómez-Barris M (2017) The extractive zone: social ecologies and decolonial perspectives. Duke University Press, Durham
Graham S, Marvin S (2001) Splintering urbanism: networked infrastructures, technological mobilities, and the urban condition. Routledge, London and New York
Harvey D (2003) The new imperialism. Oxford University Press, Oxford
Harvey D (2005) A brief history of neoliberalism. Oxford University Press, Oxford
Harvey D (2007) Neoliberalism as creative destruction. Ann Am Acad Polit Soc Sci 610:22–44
Hewitt K (1983) Interpretations of calamity. Allen & Unwin, New York
Hewitt K (1995) Excluded perspectives in the social construction of disaster. Int J Mass Emerg Disasters 13(3):317–339
Hewitt K (1997) Culture and risk. Routledge, London and New York
Heynen N (ed) (2014) The nature of cities: urban political ecology and the politics of urban metabolism. Routledge, London and New York

Heynen N, Kaika M, Swyngedouw E (eds) (2006) In the nature of cities: urban political ecology and the politics of urban metabolism. Routledge, London and New York
Imperiale J, Vanclay F (2020) Top-down reconstruction and the failure to 'build back better' resilient communities after disaster. Disaster Prev Manag 29(4):541–555
Juskus R (2023) Sacrifice zones. A genealogy and analysis of an environmental justice concept. Environ Humanit 15(1). https://doi.org/10.1215/22011919-10216129
Kaika M (2004) City of flows: modernity, nature, and the city. Routledge, London and New York
Kallis G (2021) Limits, ecomodernism and degrowth. Polit Geogr 87:102367
Katz J (2013) The big truck that went by: how the world came to save Haiti and left behind a disaster. Palgrave Macmillan, London
Keenan JM, Hill T, Gumber A (2018) Climate gentrification. From theory to empiricism in Miami-Dade County, Florida. Environ Res Lett 13(5)
Kelman I (2020) Disaster by choice: how our actions turn natural hazards into catastrophes. Oxford University Press, Oxford
Klein N (2007) The shock doctrine: the rise of disaster capitalism. Metropolitan Books, New York
Kothari A, Salleh A, Escobar A, Demaria F, Acosta A (2019) Pluriverse. A post-development dictionary. Tulika Books, New Delhi
Li TM (2014) What is land? Assembling a resource for global investment. Trans Inst Br Geogr 39(4):589–602. http://www.jstor.org/stable/24582932
Liboiron M (2021) Pollution is colonialism. Duke University Press, Durham
Loftus A (2017) Production of nature. In: International encyclopedia of geography: people, the earth, environment and technology. Wiley, Hoboken. https://doi.org/10.1002/9781118786352.wbieg0289
Macklin R (2022) Decolonizing the American literature curriculum through an environmental justice lens. Transatlantica 2 [Online]. https://doi.org/10.4000/transatlantica.20426
Malm A (2016) Fossil capital: the rise of steam power and the roots of global warming. Verso Books, London and New York
Malm A (2020) Corona, climate, chronic emergency: war communism in the twenty-first century. Verso Books, London
March H, Swyngedouw E (2022) Resilience for all or for some? Reflections through the lens of urban political ecology. In: Ruiz-Mallén I et al (eds) Urban resilience to the climate emergency. Springer, Cham
Massey D (1994) Space, place, and gender. Polity Press, Cambridge
Moore JW (2015) Capitalism in the web of life: ecology and the accumulation of capital. Verso Books, London and New York
Neumann RP (2005) Making political ecology. Oxford University Press, Oxford
Nixon R (2011) Slow violence and the environmentalism of the poor. Harvard University Press
Oliver-Smith A (2009) Nature, society, and population displacement: toward an understanding of environmental migration and social vulnerability. UNU Institute for Environment and Human Security
Peck J, Tickell A (2002) Neoliberalizing space. Antipode 34(3):380–404
Peet R, Watts M (1996) Liberation ecologies: environment, development, social movements. Routledge, London and New York
Pelling M, Dill K (2009) Disaster politics: tipping points for change in the adaptation of sociopolitical regimes. Prog Hum Geogr 34(1):21–37. https://doi.org/10.1177/0309132509105004
Pelling M, Garschagen M (2019) Put equity first in climate adaptation. Nature 569(7756):327–329
Pellow DN (2018) What is critical environmental justice? Polity Press, Cambridge
Perry RW (2018) Defining disaster: an evolving concept. In: Handbook of disaster research, pp 3–22
Perry RW, Quarantelli EL (eds) (2005) What is a disaster? New answers to old questions. Xlibris, Philadelphia, pp 311–324

Pulido L (2010) Rethinking environmental racism: white privilege and urban development in Southern California. Ann Assoc Am Geogr 90(1):12–40. https://doi.org/10.1111/0004-5608.00182
Pulido L (2016) Geographies of race and ethnicity II: environmental racism, racial capitalism and state-sanctioned violence. Prog Hum Geogr 41(4):524–533. https://doi.org/10.1177/0309132516646495. (Original work published 2017)
Pulido L (2024) Wildfire rumors and denial in the Trump era in political ecologies of the far right. In: Allen I, Ekberg K, Holgerson S, Malm A (eds). Manchester University Press, Manchester
Pyne S (2009) The human geography of fire: a research agenda. Prog Hum Geogr 33:443–446. https://doi.org/10.1177/0309132508101598
Quarantelli EL (ed) (1998) What is a disaster? Perspectives on the question. Psychology Press, London
Quarantelli EL, Boin A, Lagadec P (2017) Studying future disasters and crises: a heuristic approach. In: Handbook of disaster research. Springer, Cham, pp 61–83
Robbins P (2012) Political ecology: a critical introduction, 2nd edn. Wiley-Blackwell, Hoboken
Saurin J (2001) Global environmental crisis as the "disaster triumphant": the private capture of public goods. Environ Polit 10(4):63–84. https://doi.org/10.1080/714000578
Schuller M (2016) Humanitarian aftershocks in Haiti. Rutgers University Press, New Brunswick
Seymour R (2024) Disaster nationalism: the downfall of liberal civilization. Verso Book, London
Smith N (1984) Uneven development: nature, capital and the production of space. University of Georgia Press, Athens
Smith N (2006) There's no such thing as a natural disaster. Understanding Katrina. Item: Insights from the Social Sciences, 11 June 2006. Social Science Research Council. https://items.ssrc.org/understanding-katrina/theres-no-such-thing-as-a-natural-disaster/
Sovacool BK, Linnér BO (2016) The political economy of climate change adaptation. Palgrave Macmillan, London
Svampa M (2013) Consenso de los Commodities y lenguajes de valoración en América Latina. Nueva Sociedad [Preprint]. Available at: https://ri.conicet.gov.ar/handle/11336/6451
Svampa M (2014) ¿Territorios vacíos o territorios en disputa? Las sociedades locales, ¿entre las promesas incumplidas del desarrollo regional y el establecimiento de zonas de sacrificio? ¿Compite el avance de la explotaci´on de yacimientos no convencionales con economías regionales preexistentes? Available at: https://www.memoria.fahce.unlp.edu.ar/library?a=d&c=libros&d=Jpm5141
Svampa M (2019) Neo-extractivism dynamics in Latin America. socio-environmental conflicts, territorial turn, and new political narratives. Cambridge University Press, Cambridge USA
Swyngedouw E (2004) Social power and the urbanization of water: flows of power. Oxford University Press, Oxford
Swyngedouw E (2015) Liquid power: contested hydro-modernities in twentieth-century Spain. MIT Press, Cambridge, MA
Swyngedouw E, Ernstson H (2019) Urban political ecology in the anthropo-obscene: interruptions and possibilities. Routledge, London and New York
Szeman I, Wenzel J (2021) What do we talk about when we talk about extractivism? Text Pract 35(3):505–523. https://doi.org/10.1080/0950236X.2021.1889829
Tierney K (2021) Disasters: a sociological approach. Polity Press, Cambridge
Tuhiwai Smith L (2021) Decolonizing methodologies: research and Indigenous peoples. Zed Books, London
Valent G (2019) Disaster recovery and violence in neoliberal times: community and spatial fragmentation in L'Aquila. Disaster Prev Manag 28(4):446–459
Voyles TB (2015) Wastelanding: legacies of uranium mining in Navajo country. University of Minnesota Press, Minneapolis

Westcoat JL Jr (2015) Political ecology of risk, hazards, vulnerability, and capacities. In: Perreault T, Bridge G, McCarthy J (eds) Routledge handbook of political ecology. Routledge, London and New York, pp 293–302

Wisner B, Blaikie P, Cannon T, Davis I (1994) At risk: natural hazards, people's vulnerability, and disasters. Routledge, London and New York

Chapter 3
Ecological Disasters, Violence and Environmental Justice Movements. Case Studies from India, Italy and the US

Abstract The chapter elaborates on a broad literature on environmental violence, focusing on infrastructural and slow violence. Original case-study research conducted by one of the authors of the book is presented, with cases from India, Italy and US. The chapter is opened by an extensive exploration of methodologies for case-study research in social sciences and the usefulness of employing real-life cases to strengthen critical spatial thinking. Moreover, the different sections of the chapter combine localised empirical research and knowledge with wider theoretical debates. The case of the Sabarmati Riverfront Development Project in India focuses on infrastructural violence and shows the entanglements between the manipulation of fragile ecologies, spatial injustice, and disasters, that are even more evident in the context of the climate crisis. The case of the industrial compound in Portoscuso, Italy delves into the topic of slow ecological degradation and on the challenges of living with toxic heritage being a resource periphery. Eventually, the chapter is concluded by a short presentation of the internationally wide cases of New Orleans and the Cancer Alley, by proposing a less usual narrative re-elaboration of phenomena of slow violence and resistance promoted by environmental justice movements.

Keywords Urban political ecology · Environmental justice · Environmental history · Case-study research · Infrastructural violence · Slow violence · Sacrifice zones

3.1 Case-Study Research for Critical Spatial Thinking: A Focus on Environmental Violence

Drawing from classic texts about the use of case studies in social sciences (Yin 2009; Kreber 2001; Flyvbjerg 2001), the editors of *Planning and the Case-Study Method in Africa. The Planner in Dirty Shoes* (Duminy et al. 2014) have been emphasising the usefulness of incorporating real-life examples in urban planning teaching curricula in Africa and beyond. Such an approach would allow students to better assimilate the concepts they have been exposed to through theoretical lectures, as well as reduce

S. Armondi and G. Pessina, *Critical Spatial Thinking and Extreme Events*, PoliMI SpringerBriefs, https://doi.org/10.1007/978-3-032-18875-5_3

the distance between the classroom and real-life contexts, possibly paving the way for forms of collaboration with local actors. Ultimately, this would enhance both practical and critical skills of future urban planners and more generally, as we argue, future researchers in both the social sciences and technical disciplines.

As extensively demonstrated by the Danish economic geographer Flyvbjerg (2001, 2011), research based on case studies allows in-depth investigations of power relations and everyday practices in selected contexts and supports locally informed future policies for such settings. Therefore, the use of case studies both in research and teaching activities is particularly suited to strengthening critical spatial thinking. Although we do not have the room in this chapter to go into the details and varieties of case-study design introduced by Yin (2009), Flyvbjerg (2001) and others, we would like to emphasize that the selection "of the individual unit and the setting of its boundaries related to the choice of what is to be studied" (Flyvbjerg 2011, p. 26) is at the heart of all case-based research and teaching. Being a detailed empirical inquiry of an ongoing real-life phenomenon (Yin 2009), a multiplicity of methods for data collection are part of the researcher's toolbox, including interviews, questionnaires, participant observation, focus group discussions and other ethnographic techniques, official documents and grey literature collection, as well as spatial analysis (Duminy et al. 2014).

Quite importantly, case-study research requires an *immersion* in the space in which the considered social action is taking place (Dumont 2023) and where "local manifestations" of a certain phenomenon help the researcher to "make sense of what [is] happening" (*ibidem*, p. 451). Particularly when the researcher is new to a context and/or a topic, the preparation of the case research and the subsequent fieldwork constitute processes of *enculturation* (Schensul et al. 1999) during which it is possible to "gradually absorb the big picture" (DeWalt and DeWalt 2002, p. 68). Moreover, the presentation of case-studies to a wider audience is an opportunity for the researcher to retrace the intense emotional and bodily—in some cases traumatic—experiences associated with fieldwork (Taylor 2019). The temporal distance from the events (Paul 1953) and the use of narrative techniques (Tedlock 1991) allow field experiences to be processed and passed on to others, thus gradually sharing concepts that have been part of the *enculturation* process during the case study design and the fieldwork period. Furthermore, case study-based research also aims at contributing to the advancement of theory (Yin 2009) by challenging or downplaying popular concepts and approaches often elaborated in the so-called Global North but not always corresponding to contextualised knowledge in other geographical settings (Duminy et al. 2014).

The three cases presented in this chapter—a major water-related infrastructure and urban regeneration project in Western India; the crisis investing a highly contaminated industrial area in Sardinia, Italy; multiple disasters hitting New Orleans and the lower Mississippi area ("Cancer Alley") in Louisiana, USA—were studied by one of the authors of this book on multiple research occasions and were published or presented elsewhere as single cases (Pessina 2015, 2018, 2019, 2023, 2025), with reference to different theoretical frameworks. They are now being rewritten and presented together for the first time, adopting *environmental violence* (for a synthesis of the academic debate on the topic see Pessina 2022) as a conceptual framework and a combination of urban political ecology and environmental justice as the theoretical background (cf. Chap. 2). The three case studies are not meant to be compared, but to illustrate different dimensions of environmental violence in multiple geographical contexts, studied through different methods and with diverse temporal engagements with the field. Therefore, the way in which they are recounted as well as their length are also varying.

Based on the solid and highly influential research on violence developed by the Norwegian sociologist Johan Galtung since the Sixties (1969, 1990) in the field of peace and conflict studies, research on environmental violence has developed mostly thanks to (urban) political ecology attention to power dynamics and unbalances, as well as to environmental justice concern for inequality and ways to overcome it. Political ecologists have analysed the different forms of environmental violence according to Galtung's distinctions between direct, structural (Galtung 1969) and cultural violence (Galtung 1990). To exemplify direct and brutal forms of violence related to environmental struggles, the authors of the Environmental Justice Atlas have listed "selective assassination of environmental defenders", "unsubstantiated accusations of environmental defenders to demobilize them from their campaign", "massive coercion in a social protest" and "direct attack aiming at causing physical and psychological damage" (Navas et al. 2018, p. 653). Despite the urgency to recognise and prevent direct forms of violence against environmental defenders, in this book we decided to focus mostly on structural and not always visible forms of violence variously related to disasters, as the selected case studies will show. The definition of violence as "a concept constantly oscillating between the physical and the structural, the visible and the invisible, the natural and the social" (Pavoni and Tulumello 2020, p. 49) clearly exemplifies our approach to the topic and our choice to investigate environmental violence in its complexity.

This chapter mainly considers two dimensions of environmental violence through three disaster-related case studies: on the one hand, *infrastructural violence* (Rodgers and O'Neill 2012) and on the other hand, *slow violence* (Nixon 2011). Infrastructural violence is explored through the examination of large-scale infrastructure projects aimed at counteracting the excess or shortage of natural resources (e.g. water) or bringing development to lagging regions, while contributing to the structural marginalisation of selected populations (Luxion 2017) and irreversible impacts on ecologies on multiple scales (Swyngedouw 2015). Furthermore, the role of such infrastructure projects in the occurrence of new disasters is investigated, also considering climate change as a form of socio-ecological violence (Silver 2018). *Slow violence* as an often "spectacle deficient" (Nixon 2011, p. 47) form of "delayed destruction that is dispersed across time and space" (*ibidem*, p. 2) is mostly studied in this chapter as the outcome of water, air and soil contamination due to petrochemical production (Allen 2003; Davies 2019; Lerner 2004) or mining activities (Holterman 2014). Such form of violence is investigated with a particular attention to the unequal effects on the poorest strata of the population and to other marginal subjectivities (Nixon 2011; Pellow 2016) leading, in the most severe cases, to environmental racism (Bullard et al. 1997) and to the creation of "sacrifice zones" (Lerner

2010) where "disposable bodies" (Karim 2014) are concentrated. These structural and slow processes of violence are particularly evident when disasters and sudden climate-related events occur, as the case of Hurricane Katrina in 2005 clearly showed.

The three selected studies aim at investigating the diachronic and complex development of different typologies of disasters identifying the event (or the possibly occurring event), the response by public, private and community actors, the medium-long term consequences and changes in power dynamics on multiple scales, the social and material legacies and possible territorial fragilization processes. Less visible aspects related to phenomena that might not be traditionally identified as disasters are also considered.

3.2 Suffering Infrastructural Violence: Engineered Nature and Climate Crisis in Ahmedabad, India

In the book *The politics of climate change and uncertainty in India*, the sociologist and development studies scholar Lyla Mehta and her co-authors argue that extreme events (e.g. cyclones, floods, droughts, heatwaves etc.) create increasing uncertainty among planners, decision makers and inhabitants (Mehta et al. 2021). The choice to focus on Asia and particularly on the Indian context is explained by the authors with the high level of climate vulnerability of the country, combined with very significant socio-economic inequalities and geographical diversity. Elaborating on empirical research conducted in flood-prone coastal areas (Mumbai), drylands (the Kutch desert in Gujarat) and tropical swamp forest areas frequently hit by cyclones (the Sundarbans delta in West Bengal), the authors identify different understandings of climate-related uncertainty and responses to disasters from "above", "middle" and "below". Hence, the book investigates discourses, policies, practices and knowledge produced by a variety of actors, from the inhabitants to the decision makers.

While the contemporary intensification of extreme events makes India vulnerable, the longstanding engagement of different parts of the Indian society with ecological issues and with adaptation practices has resulted in high quality and in-depth research on the country from the perspectives of environmental history, sociology, environmental justice and territorial studies. According to the Indian environmental historian Guha (2006), environmentalist movements and academic debates in India and US have followed very different paths, while developing almost simultaneously. Even if the concept of environmental justice is relatively new in the Indian scientific debate, the Asian country has a "rich heritage of environmentalism" (Sidique and Zaidi 2024), interestingly intertwined with social justice struggles.

As Guha recalls (2006), North American environmentalism emerged around mid XIX century through the establishment of associations for the conservation of nature and wilderness (e.g. Sierra Club, Friends of the Earth) and through sport clubs (e.g. Alpine Club). Such groups of environmentalists originated in urban areas and were joined by educated middle-to-high classes. At that time and in that context, the protection of nature was not bound to any social concern nor struggle for justice. In the Sixties of the XX century health concerns started to be taken into consideration alongside with environmental issues, particularly after the publication of *Silent Spring*, the groundbreaking book by the North American biologist and writer Rachel Carson on the contamination of soil and water by pesticides (Carson 1962). Towards the beginning of the Seventies, lower classes minorities in deprived urban areas started to advocate for social justice in connection with health and environmental issues (e.g. right to medical services for all, regular waste collection etc.). Experiences such as those of the political groups of the Black Panthers or of the Young Lords Party in the Bronx (NY) were groundbreaking in this regard (Castells 1973) and raised the attention on forms of structural and environmental violence against minorities and lower classes that would have later paved the way to the debate and the actions against environmental racisms and for environmental justice (Bullard 1994).

According to some scholars, environmentalism in India can be traced back to ancient religious texts such as the Vedas and the Upanishads, where humans are presented as part of complex ecosystems made by animals, plants, as well as climatic events (Sidique and Zaidi 2024). In their view, this background informed later developments of various environmentalist movements in India, particularly since the early XX century, when the social dimension started to take centre stage in those debates. Interestingly, already in 1930 Indian scholars introduced the concept of "social ecology", according to which different phases of regional equilibrium would have been influenced by natural fluctuations (e.g. rainfall) and/or continued human actions (e.g. destruction of forests, agriculture, alteration of water systems) (Mukerjee 1930). Therefore, such early social ecologists proposed to investigate "synecologies", i.e. collective responses by different social groups to environmental changes (*ibidem*).

While in the US the first struggles for socio-environmental justice were ignited by marginalised urban groups in large cities such as New York, in India the initiative came from indigenous populations threatened by major development projects in forest and mountain areas. This was the case of the Chipko Andolan movement in the early 1970s, mostly composed by indigenous women fighting against the destruction of forests in the northern state of Uttarakhand. Or the world-famous case of the Narmada Bachao Andolan movement composed by villagers threatened by the construction of a major dam on the Narmada River, at the crossing between the states of Gujarat, Madhya Pradesh and Maharashtra. Such movement, which will be recalled later in this section, was active from the early 1960s until the mid-1990s and was joined by students, intellectuals and human right activists from the three involved states, as well as from New Delhi and other parts of the country. Several Indian scholars in social and political sciences were part of the movement and started their research on environmental conflicts, human rights, injustices and power unbalances in that context (e.g., Baviskar 1995; D'Souza 2002; Mehta 2005). As Ramachandra Guha observes in his research work conducted together with the Catalan development economist Martìnez Alier (1997), since the 1960s India has seen the emergence and the consolidation of the so-called "environmentalism of the poor". With this expression the two authors referred to struggles conducted by low-income and low-caste minorities and marginalised groups relying on their territorial resources for their livelihood.

Other forms of environmentalism emerged in the Indian subcontinent, particularly after the liberalisation of the Indian economy in 1991 and the rise of an urban middle-class in the main cities such as Delhi, Mumbai, and Calcutta. Alongside ongoing environmental struggles promoted by marginalised groups especially outside urban areas, a "bourgeois environmentalism" started appearing since the early XXI century (Baviskar 2011). The variety and richness of experiences of politicisation of urban metabolisms and environmental struggles in Indian cities (Rademacher and Sivaramakrishnan 2013) led critical urban and environmental scholars to identify that geographical context as crucial for the elaboration of an urban political ecology "beyond the West" (Zimmer 2015). In the context of the various socio-environmental conflicts that have been taking place in India in the last two decades, those related to water have been particularly frequent, as well as effective in exposing existing power unbalances and injustices in the highly unequal Indian society (Mehta 2005; Karpouzoglou and Zimmer 2012).

While Western urban scholars were starting to investigate traditional forms of coexistence between cities and water in various South Asian countries through the study of the so-called "water urbanism" (Shannon et al. 2009), Indian scholars saw major changes taking place in their cities, in the form of water-related large-scale infrastructural projects for riverfronts like the one for the Yamuna River in New Delhi (Baviskar 2011) or for seafronts as in the case of the Southern city of Chennai (Coelho and Raman 2013). According to such scholars and other critical urban planners and political ecologists, many of these projects presented some elements of *infrastructural violence*, i.e. the displacement or reduced access of a certain population to essential services due to the construction of major infrastructures (Rodgers and O'Neill 2012). As the Indian urban scholar Renu Desai observed after extensive empirical research on spatial inequalities in Ahmedabad, infrastructural violence takes place anytime "urban planning, policies and governance forge infrastructure that produces [...] inadequacies and everyday deprivations, burdens, inequities, tensions and conflicts in residents' lives" (Desai 2018, p. 89). Other scholars from India mobilised the same concept by referring to large-scale infrastructures resulting in the structural marginalisation of low-income minorities and the consolidation of already existing unequal power structures (Luxion 2017), made even more severe by increasing climate-related disasters and uncertainty (Mehta et al. 2021).

The case of the Sabarmati Riverfront Development Project in Ahmedabad, India

On July 2nd, 2023, the Indian Union Home Minister and Minister of Cooperation Amit Shah inaugurated a floating restaurant on a cruise ship in the Sabarmati River at the core of the city of Ahmedabad (Gujarat, India). As local and national newspapers reported, few days later the ship was home to a closing party with more than fifty international city mayors in the framework of India's G20 presidency. Less than a week later, flash floods hit the city as soon as the monsoon rains got more intense. Even though the event did not cause loss of lives, it heavily damaged the city. Only later, journalists found out that the water level in the Sabarmati River was increased at the beginning of July to allow the restaurant on the cruise ship to float.[1] It was not the first time that a similar event had taken place in the city, after the construction of the Sabarmati Riverfront Development Project started in 2005. In 2017, during the monsoon season, the river flooded, and the riverfront lower promenade was covered with snakes. Similar floods took place also in 2006, 2011 and 2015. Eventually, in August 2024 extremely heavy rains hit the state of Gujarat, causing the death of almost 30 people and the displacement of 24,000 inhabitants. The city of Ahmedabad was not directly hit by the disproportionate rains, but the river flooded once again.

The following paragraphs introduce the case of the Sabarmati Riverfront Development Project in Ahmedabad in the wider context of the transformations taking place in the state of Gujarat since the liberalisation of the Indian economy in the early 1990s. The case was at the core of the Ph.D. research of one of the authors of this book (Pessina 2015, 2018, 2019) and is now newly written highlighting its relevance for the debate on environmental and infrastructural violence. A major urban development, the project resulted in the manipulation of a fragile ecosystem and in increasing socio-spatial inequalities in an already heavily segregated context. Moreover, the case study shows medium to long term consequences and changes in power dynamics related to multiple shocks striking the city and making it more vulnerable to disasters. Eventually, the case study is investigated also in relation to responses to disasters, (missing) conflicts, and alternative forms of resistance.

The project at the core of the research is observed on multiple scales, ranging from the city level to the interstate development of the Narmada Canal, a major irrigation infrastructure later turned into the backbone of the industrial and urban development of the state of Gujarat. Through a combination of urban political ecology and environmental justice approaches, the case is unpacked to identify the different development phases, the actors involved, existing or new power unbalances and the dominant narratives that turned an infrastructural promise of wealth into a major socio-environmental disaster.

The interest in research on the Sabarmati Riverfront Development Project in Ahmedabad was sparked by the encounter with two very different narrations to which the author was unexpectedly exposed in 2009, while taking part in an international urban design summer school about rivers in the metropolitan area of Paris. On one hand, an enthusiastic article in a major international landscape architecture journal about the remarkable sustainability and the groundbreaking nature of such project, destined to become a model for riverfront developments in India (Scheffer and Fenk 2009). On the other hand, a master thesis in urban design analysing the ecology of the river in combination with the study of the large informal settlement which had developed in the riverbed in the last thirty years (Choudhuri 2009). After such initial interest, the research developed through a period as fellow of the Vastu Shilpa Foundation for Studies and Research in Environmental Design in Ahmedabad

---

[1] https://www.indiatoday.in/india-today-insight/story/when-a-cruise-on-sabarmati-river-flooded-ahmedabad-2405607-2023-07-12.

in 2010–2011, followed by the enrolment in the Ph.D. programme in Spatial Planning and Urban Development at Politecnico di Milano, Italy (2011–2015). Moreover, during the Ph.D. the author had the chance to spend research periods at the Institute of Development Studies (University of Sussex, UK) and at the Centre for Urban Equity (CEPT University, India).

During the first intensive fieldwork period in 2010–2011, the author conducted fifteen preliminary interviews to key actors for the Sabarmati Riverfront Development Project under the guidance of the Indian architects and urban designers Rajeev Kathpalia and Balkrishna Doshi (Vastu Shilpa Foundation). Following that, since the start of the Ph.D. she delved into an extensive literature review on socio-spatial inequalities, poverty, violence, and urban development under the guidance of the Italian sociologist Antonio Tosi. In 2012, such approach was complemented by an in-depth study of the literature on the social construction of risks (focus on scarcity/excess of water) in Asia and particularly in India under the guidance of Lyla Mehta and through the interaction with scholars at the STEPS (Social, Technological and Environmental Pathways to Sustainability) and KNOT (KNOwledge, Technology and Society) centres of the Institute for Development Studies, University of Sussex. While being a visiting Ph.D. researcher in the UK, the author had the opportunity to get exposed to the urban political ecology debate on water and power, thanks to frequent meetings with the critical geographer Erik Swyngedouw at the University of Manchester.

During the second intensive fieldwork period in Ahmedabad in 2012–2013, the author investigated a more specific literature on riverfront developments, socio-environmental conflicts and inequalities in India (focus: Ahmedabad) under the guidance of Darshini Mahadevia and Renu Desai at the Centre for Urban Equity/CEPT University. Moreover, on one hand she interviewed 35 experienced witnesses, including practitioners, public officers, engineers, urban planners, lawyers, environmental activists and journalists. On the other hand, she interviewed 55 slum dwellers affected by the construction of the riverfront. In addition to literature review and interviews, she conducted exploratory walks along the riverfront and other related water management infrastructures aimed at elaborating spatial analyses on multiple scales, and collected data, official planning/policy documents and news. The desk collection of a variety of materials related to planning choices, risk management, and dominant narratives has continued long after the completion of the Ph.D. and is still ongoing.

Major development projects against the risk of disasters? The violence of the Sabarmati Riverfront Development Project in the context of the "shock city of India"

Located in the central part of Gujarat, Ahmedabad has historically been the main commercial and economic centre of the area, being at the crossroads of major trade routes connecting the Indian subcontinent with the Arabian Sea. After the liberalisation of the Indian economy in the early 1990s, Ahmedabad found itself in a strategic location for large infrastructure developments connecting New Delhi to Mumbai, as well as for the creation of new industrial compounds, Special Economic Zones and Special Investment Regions. Therefore, Ahmedabad and the central part of Gujarat has become the backbone of what started to be defined as the "model state" of India for its economic performance since the early 2000s, especially under the guidance of the BJP nationalist party. Despite the enthusiasm of local politicians, scholars have shown how the apparent economic success of Gujarat was based on profound territorial inequalities (Kundu and Mahadevia 2002; Hirway and Mahadevia 2004), especially evident in the case of the arid western areas of Kutch, at the border with Pakistan, inhabited by tribal populations (Mehta 2005; Yagnik and Sheth 2005).

While in the last thirty years such territorial inequalities have been hardly recognised in the public discourse, the declared intention to contrast them was at the origin of major infrastructural projects for Gujarat in the post-Independence era. Already in the late fifties the Nehru government decided for the construction of a major dam along one of the sacred rivers of India, the Narmada, in a locality at the border between three Indian states (Gujarat, Maharashtra, Madhya Pradesh). Thanks to the longest canal of India, the Narmada dam, also known as Sardar Sarovar, would have provided water for irrigation to the arid region of Kutch. Scarcity was portrayed as an emergence to be tackled. As the development scholar Lyla Mehta noted after extensive research in the area with investigations of indigenous ways to deal with scarcities, water-hungry Kutch was "ironically" used by the proponents of the dam to justify the project, raising hopes and aspirations among the local populations through "powerful imagery" (Mehta 2005). In fact, the predicated amount of water never reached the area, while being captured by urban and industrial development projects in central Gujarat since the early 2000s (Luxion 2017).

Moreover, the construction of the Narmada dam sparked a decades-long environmental conflict due to the violence against the indigenous population living in the selected area and forcefully displaced without adequate compensations, as recalled by a plethora of studies from scholar-activists who joined at different times the Narmada Bachao Andolan movement (Baviskar 1995; D'Souza 2002; Gadgil and Guha 1995; Guha and Martìnez Alier 1997; Mehta 2010; Wood 2007). After twenty years of marches, pickets, legal actions and studies showing the unjust development process unfolding in relation with the construction of the dam, in the early 1990s the movement was violently repressed, especially in Gujarat, leading to the weakening of critical environmental discourses in the state and to the migration of many activists and researchers to other states or even abroad.

As we can argue through the reconstruction of the process, interviews and fieldwork (Pessina 2018), the history of the Sabarmati Riverfront Development Project in Ahmedabad, imagined in the early 1960s and remained on paper for almost forty years, is strongly related to the direct and infrastructural violence against the environmental defenders of the Narmada Bachao Andolan. A major disaster represented a turning point in the infrastructural development in Gujarat, as well as in the public response to calamities. In January 2001 a highly destructive earthquake (7.7 on the Richter scale) shook Kutch, causing almost 14,000 deaths in the area and more than 20,000 across Gujarat, with extensive damage to properties even in the city of Ahmedabad (Mishra 2012). Beyond the national and international aids to respond to the emergency, such disaster represented a turning point in the political life of

Gujarat, with effects also on what would have later become the Sabarmati Riverfront Development Project.

At that time, the Hindu nationalist BJP party held power in the state of Gujarat, at the national level, and in the city of Ahmedabad. After a period of economic, social and interreligious crises, as well as political turmoil following the closure of several textile factories in Ahmedabad ("the Manchester of India") in the 1970s, BJP had affirmed itself by playing "an exclusively rabid Hindu card in corporation/municipal elections and succeeded in winning elections in Ahmedabad, Vadodara, Surat, Rajkot, Bhavnagar, Jamnagar – the major cities in Gujarat" (Patel 2002, p. 4833) in the early 1990s and in state elections in 1995. The Kutch earthquake became a major political concern for India, as well as an opportunity for BJP politicians to gain special powers and exposure. Among them, the then national general secretary of the party, Narendra Modi, stood out of the crowd for his engagement in the monitoring and coordination of the recovery, as well as for advocating for the infrastructural and economic development of Gujarat as a response to the disaster.

It is in this framework that the construction works of the Narmada canal, supposed to bring water from the Narmada dam to the arid areas of Kutch for irrigation, were completed in 2002 and multiple initiatives for the attraction of private investments in the state were envisioned, including the International biennial summit "Vibrant Gujarat" introduced by Modi in 2003. Immediately afterwards, the works for the Sabarmati Riverfront in Ahmedabad started and accelerated in 2005, after the opening of an unplanned syphon to transfer water from the Narmada canal to the riverfront under construction, while the Narmada dam works were concluded in 2006.

The project for the Sabarmati Riverfront Development Project was proposed for the first time by the French architect Bernard Kohn in 1961 for a short stretch of the river close to the old city of Ahmedabad. The project should have been the occasion to improve the connection between the city and the river by creating new accessible public spaces. Further proposals were imagined at the end of the 1970s, when the river—dry for most of the year being non-perennial (Choudhuri 2009)—was starting to become home to numerous informal workers and dwellers living in growing slums (Breman 2004). At that time Muslim and Hindu inhabitants lived in separate slum pockets, one close to each other (Mahadevia 2007). It is only after the BJP established as the ruling party at multiple levels in 1995 that the Sabarmati riverfront project came back to Ahmedabad's municipality's agenda, through the creation of a Special Purpose Vehicle, the Sabarmati Riverfront Development Corporation. The project, proposed in 1998 by EPC/HCP, a local architecture firm and consultant, consisted in redesigning 9 km (later extended to 10.5) of the river. The main rationales behind the project included the elimination of flood hazards, the repletion of the city's aquifers, the improvement of the living conditions of the slum dwellers and the beautification of the river (EPC 1998).

The main actions included the construction of concrete walls to protect the city from floods, the narrowing of the Sabarmati riverbed in Ahmedabad to a constant width of 250 m with the reclamation of 160–200 ha of land to be partially sold to private developers, the dredging of the riverbed, the creation of a lower and upper promenade on both banks of the river and the construction of sewers. Before the actual construction of the project could take place, a further "shock" (Spodek 2011) occurred in the city of Ahmedabad, radically changing the socio-political context in which the riverfront would have developed. In February 2002, when the recovery from the Kutch earthquake was still going on, violent tensions exploded between Hindus and Muslims in Gujarat after the burning of a train carrying Hindu pilgrims in Godhra. The violence escalated and turned into what local researchers and international observers defined a *pogrom* of the Muslim population across Gujarat, with an emphasis on the rise of the Hindu ideology (*Hindutva*) and with open questions on the complicity of the state in the massacre (Bobbio 2012; Human Rights Watch 2002; Patel 2002; Yagnik and Sheth 2005; Desai 2012).

The city of Ahmedabad saw the killing of more than 1000 Muslims and the displacement of 100,000 from their homes to relief camps (Mahadevia 2007). Soon after 2002, Narendra Modi, who had already gained additional powers to respond to the Kutch emergency and to reduce interreligious tensions in Ahmedabad, became Chief Minister of Gujarat for the first time. He would have covered this role three times, before being elected Prime Minister of India (currently at his third mandate). As development scholars and urban planners have highlighted, re-establishing social order and promoting development were at the core of Modi's agenda as newly elected Chief Minister of Gujarat (Desai 2012; Luxion 2017). In this context, his first actions included the completion of the Narmada canal and the start of the Sabarmati Riverfront Development Project.

Although the project had been motivated with the reduction of flood risks and with the rehabilitation of the existing slums, the reality turned out to be different. The manipulation of the river ecology with the construction of 10 m high and 10 km long concrete walls, the dredging and reduction of the riverbed width resulted in more frequent floods downstream of Ahmedabad (Pessina 2018) and, in more recent years, in the city itself as recalled in some of the news presented in this chapter. Moreover, the project meant the displacement of around 25,000 slum dwellers and the partial relocation in state-provided social housing at a considerable distance from their previous settlements of less than half of the impacted population. As critical urban scholars have argued, this outcome implied a major infrastructural violence (Desai 2018) and increasing spatial segregation (Mahadevia 2007). Despite intense controversies around the displacement of slum dwellers has been taking place, mostly animated by human rights defenders previously active in the struggle against the Narmada dam, the project was not contested in its entirety due to the lack of an alliance among different sectors of the society, the presence of a consolidated system of power and dissent on multiple scales, especially after the response to the Kutch earthquake and the Muslim pogrom in Ahmedabad and beyond (Pessina 2018).

Moreover, the Sabarmati Riverfront Development Project, alongside with other infrastructural projects and development initiatives promoted under Modi's rule as Chief Minister of Gujarat, were sustained by narratives that proved to be appealing to the growing local (Hindu) middle class, with a combination of neoliberal development, safety, cleanliness, local identity and (Hindu) religion (Yagnik and Sheth 2005). As analysed by Renu Desai, after the 2001 Kutch earthquake and the 2002 interreligious violence resulting in the pogrom of a high number of Muslims, "Ahmedabad

[...] served a particular utility for promoting Gujarat [...]. As a gateway for pursuing investment and profit opportunities in Gujarat, Ahmedabad's re-imagining became an integral part of re-imagining and promoting Gujarat" (Desai 2012, p. 56).

The case of the Sabarmati Riverfront Development Project, briefly recalled in the previous paragraphs, shows how an apparently neutral infrastructure motivated by the reduction of risks can also be read as a striking example of environmental and infrastructural violence, if considered in the wider historical and material context in which it took place and with a combination of urban political ecology and Indian social ecology. A series of shocks and disasters have been taking place in Ahmedabad and in Gujarat in the last fifty years, in some cases related to sudden critical events and in others to longer crises. Through a selection of some of these disasters, the case study shows the entanglements between different forms of response, material, narrative and political changes. As the selection of information presented at the opening of the case study reveals, the effects of such changes, triggered by disasters, are becoming more severe and more visible in the context of the ongoing climate crisis, as critical development and urban scholars have observed in various parts of India (Mehta et al. 2021).

3.3 Living with Toxic Heritage: Slow Ecological Degradation as Disaster in Portoscuso, Italy

The second case study presented in this chapter delves mostly into less visible dimensions of environmental violence such as *slow violence* (Nixon 2011). In particular, the selected case aims at contributing to the existing international literature on the accumulation of pollutants in water, air and soil due to petrochemical production (Allen 2003; Davies 2019; Lerner 2004) or mining activities (Holterman 2014). Large part of the literature on slow but continued processes of environmental degradation (Barca 2014) developed in the US, even before the definition of *slow violence* was introduced by the South African author Nixon (2011). Early studies on the topic include those by Rachel Carson on the devastating effects of the use of chemical herbicides (Carson 1962). Following the struggles for social justice in relation with health issues promoted by lower class minorities, emerging since the beginning of the 1970s in major urban areas (see Sect. 3.2), the issue of the disproportionate burden of contamination on certain communities, especially of colour, gained a political value (Lerner 2010; Pellow 2016). It is in this context that environmental justice movements emerged from communities of colour heavily impacted by toxic contamination in the 1990s in US (Bullard et al. 1997), followed immediately afterwards by community organisations adopting similar methods and advocating for the same change in Africa and other parts of the world (see for example Ako and Olawuyi 2017; Auyero and Swistun 2009; Basu 2018).

In the framework of such studies, the unequal distribution and concentration of polluting activities in territories inhabited by minorities has been observed through a historical perspective (Barca 2014), highlighting how state and corporate choices have been enabled through time by public policies, relying on the existence of "sacrifice zones" (Lerner 2010). According to critical scholars observing phenomena of unequal contamination through such perspective, the existence of zones that could be sacrificed in the name of toxic, but strategic economic or military activities of national relevance has consolidated the control on "marginal territories" and the oppression of "marginal populations" (Barca 2014; Messerli et al. 2015), affecting the more vulnerable components of local communities (Acselrad 2004). Therefore, environmental justice scholars have been denouncing environmental racism and classism, as well as the unequal protection of communities of colour, advocating for bottom-up organisation, resistance and change (Bullard 1990, 1994).

In this line, more recent scholarship on sacrifice zones in South America has highlighted not only the descriptive dimension of the concept of *zonas de sacrificio*, showing how certain local communities are disproportionately impacted by toxic global production processes, but also its analytical potential to understand the modern colonial logic underpinning the production of spaces that are sacrificed in the name of development, progress and growth (Bolados et al. 2024) or *maldesarrollo* according to the definition introduced by the Argentinian authors Svampa and Viale (2014). Therefore, the concept of *zonas de sacrificio* has been adopted by various anti-extractivist struggles in Latin America denouncing the destruction of ecosystems and other forms of environmental violence (Bolados et al. 2024).

As recalled by the Dutch ecofeminist and political scientist Maite Smet in her studies on Chilean community movements active in *zonas de sacrificio* (Smet 2024), an increasing number of experiences of resignification and attempts to overcome the stigma associated to living in such areas can be observed, as in the case of the collective Mujeres por el Buen Vivir (MUBUVI) in the Quintero-Puchuncaví bay. The area, already studied as an example of the paradoxical phenomenon of "noxious deindustrialisation" by the Italian and Chilean sociologists Feltrin and Julio Medel (2024), has been home to major environmental justice movements lead by women in the past such as the Mujeres de zona de sacrificio Quintero-Puchuncaví. Therefore, such example, alongside with others in Latin America shows an interesting legacy of previous movements and the capacity by local communities to incorporate different forms of struggle and approaches, from the environmental justice repertoire to the *buen vivir* practices (see Chap. 4).

Such approaches are grounded in the bodily experience of local communities and particularly of women, as cases in different parts of the world are showing. Slow violence has been defined as "spectacle deficient" (Nixon 2011, p. 47), but signals of contamination can be captured through careful attention to bodily reactions and the insurgence of forms of illness as a result of daily prolonged exposure to toxicity. Therefore, the British political and environmental geographer Thom Davies has been advocating for slow observations, opposing the idea that toxic geographies are not visible, while proposing to ask who the toxicity is "out of sight" (Davies 2019).

Such approach, and the pivotal role of local women in noticing the contamination and advocating for change, has been observed also in several other context, including Italy, with in-depth research on illegal toxic waste dumping in "Terra dei

Fuochi" (Armiero and De Rosa 2016; Iengo and Armiero 2017; De Rosa 2018), highly polluting metallurgical industrial activities in Taranto (Barca and Leonardi 2018) or petrochemical industry in Gela (Privitera 2024), just to mention some of the best known examples. Such studies combine international and national phenomena (e.g. economic restructuring, illicit trade, and delocalisation of multinational companies) with in-depth studies of the selected territories, with a focus on the affected communities, their capacity (and difficulties) to organise themselves as a movement against environmental injustices and for reparation.

In addition to studies conducted by political ecologists, cases of slow violence in Italy are increasingly present also in the environmental history scholarship (see for example Adorno 2023; Caruso 2019; Ferrarese 2019; Mioche and Bertilorenzi 2015), thus contributing to the international debate among historians on toxicity and environmental violence in contemporary territories (Bécot and La Naour 2023). Given the fundamental relation between space and time characterising slow violence, emerging cross-country collaborations between urban planners and environmental historians are becoming more frequent (Pessina and Brenu 2022). Moreover, multidisciplinary studies including conservation architects are increasingly tackling the issue and the materiality of "toxic legacies" (Bangstad and Peturdottir 2022; Brenu et al. 2024).

The case study presented in this section—the industrial area of Portoscuso, in Sardinia (Italy)—elaborates on the international and Italian debate on slow violence through a combination of multiple perspectives including critical urban studies, urban political ecology, environmental justice, and environmental history. Differently from the Indian case presented in the previous section, this Sardinian context was not the object of dedicated research since the beginning but was encountered by the author in the framework of a consultancy and later developed as a case study. As recalled in a dedicated article focusing on ethnographies of energy production in times of transition (Pessina 2023), the author worked on an atypical contract as self-employed academic consultant for two distinct periods of 4–6 months between 2016 and 2018. The object of the consultancy was a project for the decommissioning and sale of more than twenty thermoelectric power stations and a former lignite mine that were considered marginal by the multinational utility—the former Italian energy state company—owning them (Geroldi and Pessina 2018).

The author worked alongside a landscape architect on five power stations located in different parts of Italy, as well as on the lignite mine. After rapid surveys and intense visits condensed in one or two days, their main task was to quickly elaborate territorial and architectural reports on the power stations. Being frustrated by the speed of the work, the standardised output, as well as by other dimensions common to precarious workers active in knowledge fields, such as economic and contractual insecurity, skills mismatch and the difficulty to grasp the wider context of the work (Murgia 2015), the author started noting down field site observations and dialogues with the workers. A new "kind of attention" (Poulos 2012, p. 67) was developed, while "stumbling into possibilities" (*ibidem*, p. 66) offered by the fieldwork. The author hence started to collect and later transform "unplanned moments in the field into data" (Fuji 2015, p. 525). Such approach is common to other experiences of "accidental ethnography" (Poulos 2012; Fuji 2015), i.e. "a methodology for practitioners to examine past experiences and contribute their findings to scholarly discourse" (Levitan et al. 2020, p. 336).

The Portoscuso oil-fuelled power station in Sardinia, located in the Portovesme industrial compound, was the sixth and last plant surveyed by the author before quitting the job as academic consultant. At the time of that job, the author could spend one day at the Regional Department of Industry of Sardinia (October 2017) where she met institutional stakeholders. One month later, she was granted four days in Portoscuso and Cagliari to visit the closed power station and conduct interviews with local stakeholders and key informants, including the mayor, the vice-mayor responsible for local environmental affairs, a municipal officer, the captain harbourmaster, two unionists, the president of the Sardinia Geopark, the head of the Carbonia-Iglesias province Industrial Consortium, the president of the Regional Scientific Committee of the Legambiente environmental association, the Sulcis Extraordinary Plan's coordinator, the Regional Confederation of Industries' president, and the Regional councillor for Industry. The author was always accompanied by a representative of the multinational utility for which she was working and was not allowed to talk to industrialists and workers of other companies of Portovesme compound nor to the most radical environmentalists.

After having quitted the job as consultant, the author continued to be engaged with the case study, expanding the interest to the whole Portovesme industrial areas and to the wider Sulcis former mining region, while working as postdoc (2019–2021) and later untenured assistant professor (2021–now) at the Department of Architecture and Urban Studies, Politecnico di Milano. Elaborating on debates in the fields of economic geography, critical urban studies and anthropology on "resource peripheries" (Hayter et al. 2003; Irarrázaval and Arias-Loyola 2021), "hinterlands of the Capitalocene" (Brenner and Katsikis 2020), marginal territories (Keil 2020), and "laborless landscapes" (Franquesa 2022), the author proposes an investigation of a "resource periphery/waste core" (Pessina 2025) in Italy targeted by the EU Just Transition Fund 2021–2027 and recently included among the EU Strategic Projects under the European Critical Raw Materials Act 2024.

As observed by economic geographers investigating resource peripheries since the 1990s, such territories are characterised by the presence of resources that are only partially processed on site, before being commercialised in other places that will gain from the higher value share of the final product (Hayter et al. 2003). In such areas, scholars observed "a clash of industrial, environmental, cultural and geopolitical dimensions not found in cores", as well as forms of industrialisation that often create "instability, crisis, and dependence" (*ibidem*, p. 19). In this framework, mining territories are a specific kind of resource periphery necessary to sustain global cities and core areas (Arboleda 2020; Brenner and Katsikis 2020; Geroldi and Pessina 2024), while being external to the urban continuum and the policies effectively governing it (Keil 2020). Moreover, resource peripheries are often experiencing paradoxical processes of "noxious deindustrialisation" (Feltrin and Julio Medel 2024), i.e. increasing unemployment in areas where highly polluting areas are still (partially) operating in the framework of the "international division of labour and noxiousness" (*ibidem*, p. 173). Hence, it is not unusual to observe that resource peripheries could be also considered as sacrifice zones (Lerner 2010; Bolados et al. 2024), as in the case of the Portoscuso industrial area in Sardinia.

The area, once a small fishing village in the southwestern tip of Sardinia and later the port of the Sulcis mining region, turned into a major aluminium production pole as a result of public policies on multiple scales, aimed at contrasting the crisis of coal mining in the region, after WWII, and bringing economic development to southern Italian regions (Mioche and Bertilorenzi 2015). In Sardinia, regional, national and supranational funds were mobilised with the declared aim of sustaining small to medium enterprises using local resources and creating value to be shared onsite. Nevertheless, most of the funds were captured by large petrochemical and metallurgical companies from more economically developed Northern regions in Italy or from abroad (Pessina and Brenu 2022). Hence, large industrial areas emerged in different parts of the island, including a major petrochemical pole in Porto Torres, a strategic oil refining site for the whole Mediterranean in Sarroch, a large chemical compound on the outskirts of Cagliari, and what later became the only Italian active aluminium production area in Portoscuso.

Taking advantage of financial incentives and of the creation of a massive coal-fuelled power station ("Supercentrale") able to sustain energy-intensive activities, multiple metallurgical industries settled in the 700 ha Portovesme area between 1966 and 1968. A refinery extracting alumina from bauxite (Eurallumina) was a joint venture between one of the Italian state authorities and the Kaiser American company. An alumina-to-primary aluminium plant had been created through an agreement between Montecatini national company and the Italian government and included an oil-fuelled power plant. A zinc and lead smelting plant was created with state and private funds. Most of such companies relied on highly polluting chemical processes that became evident when the toxic red muds resulting from the bauxite-to-alumina refining stopped being discharged in the sea, due to protests by the local tuna fishermen and started to accumulate in a basin between the industrial pole and the coast.

Soon after their establishment, most of the companies experienced multiple changes of properties becoming owned almost exclusively by multinational entities (e.g. ALCOA, Rusal, Glencore, ENI). In the early 1990s, while the issue of contamination was beginning to be acknowledged both locally and nationally, the industries started a long path towards the crisis resulting in the almost complete suspension or closure of activities between 2008 and 2012. Despite having been already included among the Italian "high environmental risk areas" (1990) and those of "national interest for remediation" (2001), the recognition of contamination was not enough to protect the local population. Therefore, an ordinance for the "prohibition of marketing or distribution in any way" of local products including wine, grapes, olives, tomatoes was issued in 2014 by the mayor, while blood tests of children were showing the presence of lead in their bodies. Within two years, the area was also registered among the "complex industrial crisis areas" of Italy, destined to receive extraordinary funds to contrast the decline.

Similarly to other areas undergoing processes of "noxious deindustrialisation" (Feltrin and Julio Medel 2024) in the world, Portoscuso has seen the permanence of the toxicity and the display of slow violence through time and across space, while rampant unemployment was taking place. Unlike other cases in which vocal forms of environmental protection developed within and around polluting factories, also through the contribution of workers and unionists (Barca and Leonardi 2018), the "difficult balance" (Conte and Ferrarese 2020) between labour and health remained very tense and environmental protection movements did not form locally (Pessina 2023). While the area is recognised as a sacrifice zone with a massive toxic heritage by environmental defenders living in other parts of the island and of Italy, only a small fraction of the local population including doctors and schoolteachers is using such concept to define the conditions they are living in (Pessina 2025).

Such absence of an organised and vocal movement can be also related to the very small dimension (less than 5000 inhabitants) and relative isolation of the town of Portoscuso, as well as to the disproportionate historical presence of major extractive companies, as observed for other former mining regions and resource peripheries in various parts of the world (Hayter et al. 2003). While the local society is highly fragmented, major European fundings are investing the region, as the Sulcis former mining area was selected as one of the two Italian territories to benefit from the Just Transition Fund programme within the EU Green Deal (Greco et al. 2023). Nowadays, the most concerned local citizens are wondering whether this will turn into another promise of occupation in polluting industries. The EU decision to include among the Strategic Projects for the Critical Raw Materials, Act 2024 the conversion of one of the metallurgical plants in crisis into lithium black mass refinery seems to support the worries of the concerned citizens. Still, a wider alliance with sacrifice zone across Sardinia, Italy and other parts of the world might support a real change and an effective transition to just and less harmful economies.

3.4 Responding to Environmental Discrimination and Disasters: Environmental Justice Movements in New Orleans and the Cancer Alley, USA

As already recalled in Chap. 1 and in various sections of the present chapter, it is crucial to understand the unfolding of a disaster through space and time, given that the policies, actions, and priorities to recover from a shock can cause further—less visible, yet irreversible—disasters. As the study of some exemplary cases such as the Katrina hurricane in New Orleans have shown, disasters highlight existing socio-spatial inequalities, power unbalances and socio-political regimes (Pelling and Dill 2009).

Hence, it should come as no surprise that the Scottish Marxist geographer Neil Smith observed the effects of Katrina and declared that "there's no such thing as a natural disaster" (Smith 2006). Given the scale of the hurricane and the impact on the city and the society, national and international media attention to the event was extremely high in the immediate aftermath. Yet, the disaster and response to it continued to preoccupy international scholars dealing with critical disaster studies and residents exposed to the effects of the recovery and increasing extreme events. Through the elaboration of some interviews to experienced witnesses among residents and activists collected by the author of the present chapter in 2019, the following paragraphs show how the effects of the 2005 disaster were still a source of concern and a trigger for further socio-environmental struggles, even 15 years after the event. The research was conducted by the author in the framework of the intensive *Anthropocene River Campus* in November 2019, which was jointly promoted by the Max Planck Institute for the History of Science and by the New Orleans Centre for Gulf South at Tulane University. Differently from what has been written about India and Sardinia in the previous sections of the chapter, the short New Orleans/Cancer Alley case studies contribute to slow environmental violence research through a narrative way to recall contamination, disasters and forms of resistance.

As Beth Butler, member of the ACORN association declared, speaking about the unequal effects of the disaster, the most heavily impacted neighbourhoods were those in the low-lying areas of the city of New Orleans, mostly inhabited by black residents, like Gentilly (4–5 m below the sea level). In her words: "During Katrina, Gentilly was covered with water for weeks on end and then there was mud, black mud that seemed to have entered the bones of those who survived" (Pessina 2022, p. 95). While highlighting the extended temporality of the disaster that was spanning beyond the days in which the hurricane hit the city, Butler underlined the impact on an already segregated and underprivileged neighbourhood that was radically changed afterwards, especially in its population. As she observed, "after Katrina, some houses were rebuilt and bought by 'super-owners' who rent them to tourists – nearly all white - on Airbnb" (*ibidem*).

It is in this context that her organisation started a struggle on the side of the marginalised inhabitants—almost all low-income and black—in some neighbourhoods of the city, with a constant presence in the Lower Ninth Ward. The area is in the eastern part of the city and is close to the industrial canal—renamed as "Hurricane Superhighway" by climatologists—connecting the Lake Pontchartrain to the Mississippi river for the transfer of goods by cargo ships. The Lower Ninth Ward was among those most affected by Katrina because the banks of the canal broke down and the water submerged it entirely. As Beth Butler recalled, in front of the damaged houses stood a red sign that read 'Save our neighbourhood. NO BULLDOZING' with the phone number of the ACORN organisation, which was fighting against the demolitions and the displacement of the black low-income owners. The struggle to avoid real estate speculations and "ethnic cleansing", as she defined the phenomenon taking place there and in other neighbourhoods of the city, continued for months and was partially successful.

Other areas of the city were completely cleared after the hurricane, like parts of the Tremé neighbourhood. It is the case of the Lafitte Projects, a public housing complex created in the Forties to host 900 low-income black families. Following Katrina, they were emptied, partly demolished and fenced off, although the hurricane had not damaged them, and they were still in good conditions. During a walk through the area, Griffin read a statement by a local republican less than one month after the hurricane: "Finally, we cleaned up the housing projects in New Orleans. We couldn't do it, but God could" (*Griffin n.d.*, p. 97). In the subsequent months, New Orleans lost one hundred thousand blacks out of a total population of five hundred thousand. Some died, but most were 'relocated' to other cities in Louisiana or to other states.

According to local activists and researchers, after the emergency, attention had waned on the effects of Katrina, but these had worsened over time. As Beth Butler declared: "The sewers have stopped working decently, electricity blows every time there is a bit of wind, sanitation is collapsing and schools worse. It turns out there's lead in the water pipes. Lead, can you imagine?" Despite the situation, she and her organisation, like many others, continue to stay in New Orleans and to animate local struggles. Thinking about the increasing frequency of extreme events and the dire state of the city, despite the investments in recovery and resilience plans, she concluded: "If another hurricane like Katrina comes, there won't be much we can do. Maybe it will flood us forever. Maybe we are waiting for it. In the meantime, we continue to fight against injustice" (*ibidem*, p. 96).

The short period of research conducted by the author in the framework of the *Anthropocene River Campus* covered not only the urban area of New Orleans but also the surrounding region, and particularly the portion of the Mississippi River north of the city infamously known as the "Cancer alley" or "Toxic Alley" (Allen 2003). Such areas, where a disproportionate concentration of polluting factories in the chemical and metallurgical sectors were established around low-income black settlements, have been the object of the pioneering North American studies on environmental justice (e.g. Bullard et al. 1997) and are still home to multiple practices of resistance to "slow violence" (Davies 2019), including citizen data collection and activism through art (Pessina 2022).

New Orleans and the Cancer Alley are recognised among critical disaster scholars, environmental sociologists and political ecologists as landmark case studies to understand the complexity and the socio-political dimensions of disasters which unfold over time and across geographical scales. The increasing frequency and violence of extreme events such as hurricanes and storms hitting the area made those cases closely observed by climatologists and environmental justice scholars alike.

References

Acselrad H (2004) De "bota-foras" e "zonas de sacrificio" um panorama dos conflitos ambientais no Estado do Rio de Janeiro. In: Acselrad H (ed) Conflito Social e Meio Ambiente no Estado do Rio de Janeiro. Relume Dumará, Rio de Janeiro, pp 7–18

Adorno S (2023) La bonifica dei siti di interesse nazionale: il caso della Rada di Augusta. Meridiana 108(3):101–127

Ako RT, Olawuyi DS (2017) Environmental justice in Nigeria: divergent tales, paradoxes and future prospects. In: Holifield R, Chakraborty J, Walker G (eds) The Routledge handbook of environmental justice. Routledge, London and New York, pp 567–577

Allen B (2003) Uneasy alchemy citizens and experts in Louisiana's chemical corridor disputes. MIT Press, Cambridge, MA

Arboleda M (2020) Planetary mine: territories of extraction under late capitalism. Verso, London

Armiero M, De Rosa SP (2016) Political effluvia: smells, revelations, and the politicization of daily experience in Naples, Italy. In: Thorpe J, Rutherford S, Sandberg LA (eds) Methodological challenges in nature-culture and environmental history research. Routledge, London, pp 193–206

Auyero J, Swistun DA (2009) Flammable: environmental suffering in an Argentinian Shantytown. Oxford University Press, Oxford and New York

Bangstad TR, Peturdottir P (eds) (2022) Heritage ecologies. Routledge, London and New York

Barca S (2014) Telling the right story: environmental violence and liberation narratives. Environ Hist 20(4):535–546

Barca S, Leonardi E (2018) Working-class ecology and union politics: a conceptual topology. Globalizations 15(4):487–503

Basu P (2018) Environmental justice in South and Southeast Asia: inequalities and struggles in rural and urban contexts. In: Holifield R, Chakraborty J, Walker G (eds) The Routledge handbook of environmental justice. Routledge, New York, pp 603–614

Baviskar A (1995) In the belly of the river: tribal conflicts over development in the Narmada Valley. Oxford University Press, New Delhi

Baviskar A (2011) What the eye does not see: the Yamuna in the imagination of Delhi. Econ Polit Wkly 46(50):45–53

Bécot R, La Naour G (2023) Vivre et lutter dans un monde toxique. Violence environnementale et santé à l'âge du pétrole. Paris, Seoul

Bobbio T (2012) Making Gujarat vibrant: Hindutva, development and the rise of subnationalism in India. Third World Q 33(4):657–672

Bolados P, Espinoza L, Panez A (2024) Una aproximación decolonial a las Zonas de Sacrificio en Chile. In: Dossier: Neoextractivismo, crisis climática y desastres: Miradas desde Argentina, Brasil y Chile 21. Vibrant. Virtual Brazilian Anthropology. https://doi.org/10.1590/1809-43412024v21d603

Breman J (2004) The making and unmaking of an industrial working class: sliding down the labour hierarchy in Ahmedabad, India. Amsterdam University Press, Amsterdam

Brenner N, Katsikis N (2020) Operational landscapes: hinterlands of the Capitalocene. Landscapist 90(1):22–31

Brenu N, Pessina G, Tiganea O (2024) Toxic legacies. Officina 47:6–79

Bullard RD (1990) Dumping in Dixie: race, class, and environmental quality. Westview Press, Boulder, CO

Bullard RD (1994) Unequal protection: environmental justice and communities of color. Random House, New York

Bullard RD, Johnson GS, Wright BH (1997) Confronting environmental injustice: it's the right thing to do. Race Gender Class 5(1):63–79

Carson R (1962) Silent spring. Houghton Mifflin, Boston, MA

Caruso V (2019) Territorio e deindustrializzazione. Gli anni Settanta e le origini del declino economico di Napoli est. Meridiana 96:209–230

Castells M (1973) Luttes Urbaines. Maspero, Paris

Choudhuri P (2009) Re-structuring the development along a non-perennial river. Case: Sabarmati River, Ahmedabad. Master thesis, CEPT University, Ahmedabad (India)

Coelho K, Raman N (2013) From the frying pan to the floodplain: negotiating land, water, and fire in Chennai's development. In: Rademacher AM, Sivaramakrishnan K (eds) Ecologies of urbanism in India: metropolitan civility and sustainability. Hong Kong University Press, Hong Kong, pp 145–168

Conte A, Ferrarese G (eds) (2020) Un bilanciamento difficile: industria e ambiente dal dopoguerra a oggi. Le Penseur, Brienza

Davies T (2019) Slow violence and toxic geographies: 'out of sight' to whom? Environ Plan C Polit Space 40(2):409–427
De Rosa SP (2018) A political geography of 'waste wars' in Campania (Italy): competing territorialisations and socio-environmental conflicts. Polit Geogr 67:46–55
Desai R (2012) Entrepreneurial urbanism in the time of the Hindutva. City imagineering, place marketing, and citizenship in Ahmedabad. In: Desai R, Sanyal R (eds) Urbanizing citizenship. Contested spaces in Indian cities. Sage, London, pp 31–57
Desai R (2018) Urban planning, water provisioning and infrastructural violence at public housing resettlement sites in Ahmedabad, India. Water Altern 11:86–105
DeWalt KM, DeWalt BR (2002) Participant observation: a guide for fieldworkers. AltaMira Press, Walnut Creek, CA
D'Souza D (2002) The Narmada dammed: an inquiry into the politics of development. Penguin Books India, New Delhi
Duminy J, Andreasen J, Lerise F, Odendaal N, Watson V (eds) (2014) Planning and the case-study method in Africa. The planner in dirty shoes. Palgrave Macmillan, London
Dumont G (2023) Immersion in organisational ethnography. Four methodological requirements to immerse oneself in the field. Organ Res Methods 26(3):441–458
EPC (1998) Proposal for the Sabarmati riverfront development. Environmental Planning Collaborative, Ahmedabad
Ferrarese G (2019) L'Anic di Pisticci. La parabola dell'industria di Stato sull'osso del Mezzogiorno. Meridiana 95:247–268
Flyvbjerg B (2001) Making social science matter: why social inquiry fails and how it can succeed again. Cambridge University Press, Cambridge
Feltrin L, Julio Medel G (2024) Noxious deindustrialisation and extractivism: Quintero-Puchuncaví in the international division of labour and noxiousness, New Political Economy. 29(2): 173–191
Flyvbjerg B (2011) Case study. In: Denzin NK, Lincoln YS (eds) The Sage handbook of qualitative research. Sage, Thousand Oaks, CA, pp 301–316
Franquesa J (2022) Unruly workers and laborless landscapes. The role of marginal places and redundant people in energy transitions. In: Kasimir S, Gill L (eds) The Routledge handbook of the anthropology of labour. Routledge, New York and London
Fuji LA (2015) Five stories of accidental ethnography. Turning unplanned moments in the field into data. Qual Res 15(4):525–539
Gadgil M, Guha R (1995) Ecology and equity: the use and abuse of nature in contemporary India. Penguin, New Delhi
Galtung J (1969) Violence, peace and peace research. J Peace Res 6(3):167–191
Galtung J (1990) Cultural Violence, Journal of Peace Research. 27(3): 291–305
Geroldi C, Pessina G (2018) Le centrali elettriche coinvolte nel progetto Futur-e. Territorio 86(3):72–85
Geroldi C, Pessina G (2024) Reading mining energy landscapes, over time. In: Calabrò F, Madureira L, Morabito FC, Piñeira Mantiñán MJ (eds) Networks, markets & people: communities, institutions and enterprises towards post-humanism epistemologies and AI challenges, vol 1. Springer, Cham, pp 65–74
Greco L, Perra S, Pruna ML (2023) La sfida della 'transizione giusta' in due aree di antica industrializzazione. Meridiana 106:39–68
Griffin SM (n.d.) Displaced. https://www.displacedneworleans.com/
Guha R (2006) How much should a person consume? Environmentalism in India and the United States. University of California Press, Los Angeles
Guha R, Martìnez Alier J (1997) Varieties of environmentalism. Essays north and south. Routledge, London
Hayter R, Barnes TJ, Bradshaw MJ (2003) Relocating resource peripheries to the core of economic geography's theorizing: rationale and agenda. Area 35(1):15–23
Hirway I, Mahadevia D (2004) Gujarat human development report-2004 completed. CFDA Centre for Development Alternatives

Holterman D (2014) Slow violence, extraction and human rights defense in Tanzania: notes from the field. Resour Policy 40(C):59–65
Human Rights Watch (2002) "We have no order to save you": state participation and complicity in communal violence in Gujarat. Human Rights Watch, New York
Iengo I, Armiero M (2017) The politicization of ill bodies in Campania Italy, Journal of Political Ecology. 24(1): 44–58
Irarrázaval F, Arias-Loyola M (2021) Resource peripheries in the global economy. Networks, scales and places of extraction. Springer, Cham
Karim L (2014) Garment factory catastrophe and feminist practices in Bangladesh. Anthropol Now 6(1):52–63
Karpouzoglou T, Zimmer A (2012) Closing the gap between 'expert' and 'lay' knowledge in the governance of wastewater: lessons and reflections from New Delhi. IDS Bull 43(2):59–68
Keil R (2020) An urban political ecology for a world of cities. Urban Stud 20(57):2357–2370
Kundu A, Mahadevia D (2002) Poverty and vulnerability in a globalising metropolis: Ahmedabad. Manak Publications Pvt Limited, Ahmedabad
Kreber C (2001) Learning Experientially through Case Studies? A Conceptual Analysis Teaching in Higher Education 6(2): 217–228
Lerner S (2004) Diamond: a struggle for environmental justice in Louisiana chemical corridor. MIT Press, Cambridge, MA
Lerner S (2010) Sacrifice zones: the front lines of toxic chemical exposure in the United States. MIT Press, Cambridge, MA
Levitan J, Carr-Chellman D, Carr-Chellman A (2020) Accidental ethnography: a method for practitioner-based education research. Action Res 18(3):336–352
Luxion M (2017) Nation-building, industrialisation, and spectacle. Political functions of Gujarat's Narmada pipeline project. Water Altern 10(2):208–232
Mahadevia D (2007) A city with many borders: beyond ghettoisation in Ahmedabad. In: Shaw A (ed) Indian cities in transition. Orient Longman, Hyderabad, pp 341–389
Mehta L (2005) The politics and poetics of water: the naturalisation of scarcity in western India. Orient Longman, New Delhi
Mehta MG (2010) A river of no dissent: Narmada movement and coercive Gujarati nativism. South Asian Hist Cult 1(4):509–528
Mehta L, Adam HN, Srivastava S (2021) The politics of climate change and uncertainty in India. Routledge, London
Messerli P, Peters A, Schoenweger O, Nanhthavong V, Heinimann A (2015) Marginal lands or marginal people? Analysing key processes determining the outcomes of large-scale land acquisitions in Lao PDR and Cambodia. Int Dev Policy. http://journals.openedition.org/poldev/2037
Mioche P, Bertilorenzi M (2015) Le résidus de l'alumine à Portovesme (Italie) et à Gardanne/Cassis (France) des années 1960 à nos jours. In: Centemeri L, Daumalin X (eds) Pollutions industrielles et espaces méditerranéens (XVIII–XXI siècle). Karthala-MMSH, Aix-en-Provence, pp 275–299
Mishra PK (2012) The Kutch earthquake 2001. National Institute of Disaster Management
Mukerjee R (1930) An ecological approach to sociology. Sociol Rev 22(4):281–291
Murgia A (2015) Obverse and reverse sides of precariousness: young highly educated workers between passions and skill mismatch. Soc Altern 34(4):14–21
Navas G, Mingorría S, Aguilar B (2018) Violence in environmental conflicts: the need for a multidimensional approach. Sustain Sci 13(4):649–660
Nixon R (2011) Slow violence and the environmentalism of the poor. Harvard University Press, Cambridge MA, London
Patel G (2002) Narendra Modi's one-day cricket match. Econ Polit Wkly 37(48):4826–4837
Paul BD (1953) Interview techniques and field relationships. In: Kroeber AL (ed) Anthropology today. University of Chicago Press, Chicago, pp 430–451
Pavoni A, Tulumello S (2020) What is urban violence? Prog Hum Geogr 44(1):49–76

Pelling M, Dill K (2009) Disaster politics: tipping points for change in the adaptation of sociopolitical regimes. Prog Hum Geogr 34(1):21–37

Pellow DN (2016) Toward critical environmental justice studies: black lives matter as an environmental justice challenge. Du Bois Rev Soc Sci Res Race 13(2):221–236

Pessina G (2015) The necessary risk: water, fear and development in western India. The case of the Sabarmati riverfront development project. Doctoral dissertation, Ph.D. programme in spatial planning and urban development, Politecnico di Milano, 26th cycle

Pessina G (2018) The 'missing conflict' of the Sabarmati riverfront. authoritarian governance, neoliberalism, and water in Ahmedabad. Partecip Confl 11(3):692–714

Pessina G (2019) The brand-new riverfront and the historical center: narratives and open questions in contemporary Ahmedabad, India. Eur J Creat Pract Cities Landsc 2(1):65–92

Pessina G (2022) Come querce negli uragani. 'Disastri e resistenza' a New Orleans e dintorni. In: Rossomando L et al (eds), vol 5. Lo Stato delle Città, pp 93–99

Pessina G (2023) Accidental ethnographies of energy transition. A project to decarbonise Italy's power stations as recounted by an urban planning consultant. Etnogr Ric Qual 2:247–272

Pessina G (2025) Resource peripheries/waste cores: labour regimes and 'just transition' in South Sardinia. Presentation at the 43rd international labour process conference—labour process and the environmental crisis in a multipolar world, Universidad Alberto Hurtado, Santiago de Chile, 23–25 Apr 2025

Pessina G, Brenu N (2022) Una promessa dopo l'altra: dal petrolchimico alla crescita verde di Porto Torres? In: Fior M, Kërçuku A, Mattioli C, Pessina G (eds) Fragilità nei Territori della Produzione. FrancoAngeli, Milano, pp 217–235

Poulos CN (2012) Accidental ethnography: an inquiry into family secrecy. Left Coast Press, Walnut Creek, CA

Privitera E (2024) (Counter)mapping toxic legacies. A small, slow, and street approach in the risk landscapes of Gela. Officina 47:41–49

Rademacher A, Sivaramakrishnan K (2013) Ecologies of urbanism in India. Hong Kong University Press, Hong Kong

Rodgers D, O'Neill B (2012) Infrastructural violence: introduction to the special issue. Ethnography 13(4):401–412

Scheffer K, Fenk A (2009) Sabarmati riverfront development project. Topos 68:42–45

Schensul SL, Schensul JJ, LeCompte MD (1999) Essential ethnographic methods. Observations, interviews, questionnaires. AltaMira, Walnut Creek, CA

Shannon K, De Meulder B, d'Auria V, Gosseye J (eds) (2009) Water urbanisms. SUN, Amsterdam

Sidique U, Zaidi K (2024) Rising to the challenge: a systematic review of the development of environmental justice in India. India Rev 23(4):325–344

Silver J (2018) Suffocating cities. Climate change as socio-ecological violence. In: Ernstson H, Swyngedouw E (eds) Urban political ecology in the anthropo-obscene. Routledge, London, pp 129–147

Smet M (2024) Resignificando una 'zona de sacrificio'. Entrelazando visiones regenerativas, ecofeministas y del buen vivir. Regenerativa. Published online on 16 Apr 2024. https://www.regenerativa.cl/post/resignificando-una-zona-de-sacrificio-entrelazando-visiones-regenerativas-ecofeministas-y-del-bu

Smith N (2006) There's no such thing as a natural disaster. Insights from the Social Sciences, 11 June 2006. https://items.ssrc.org/understanding-katrina/theres-no-such-thing-as-a-natural-disaster/

Spodek H (2011) Ahmedabad. Shock city of twentieth century India. Indiana University Press, Bloomington

Svampa M, Viale E (2014) Maldesarrollo. La Argentina del extractivismo y el despojo. Katz, Buenos Aires

Swyngedouw E (2015) Liquid power: contested hydro-modernities in twentieth-century Spain. MIT Press, Boston

Taylor S (2019) The long shadows cast by the field: violence, trauma, and the ethnographic researcher. Fennia 197(2):183–199

Tedlock B (1991) From participant observation to the observation of participation: the emergence of narrative ethnography. J Anthropol Res 47(1):69–94
Wood JR (2007) The politics of water resource development in India: the case of Narmada. Sage, New Delhi
Yagnik A, Sheth (2005) The shaping of modern Gujarat. Plurality, Hindutva and beyond. New Penguin Books, Delhi
Yin RK (2009) Case study research: design and methods, 4th edn. Sage, Thousand Oaks, CA
Zimmer A (2015) Urban political ecology 'beyond the west': engaging with South Asian urban studies. In: Bryant R (ed) International handbook of political ecology. Edward Elgar, Cheltenham

Chapter 4
Crises and Extreme Events as a Portal for Reimagining Other Futures

Abstract The chapter aims at stimulating imagination in the aftermath of disasters, a crucial step in developing alternative responses. Extreme events disrupt standard planning systems, forcing local communities to experiment with makeshift survival solutions and infrastructures. On one hand, they expose the failures of current systems; on the other, they open the way for new ideas, policies, and practices. The prevailing approach to disaster preparedness is based on risk management, technocratic solutions, and forecasting. However, critical disaster studies argue that preparedness for extreme events and post-disaster management should extend beyond technical emergency measures to include rethinking and reimagining social and spatial structures within a polycrisis (ecological, social, economic and political) setting, in order not to repeat, exacerbate, or reproduce existing vulnerabilities. Furthermore, instead of viewing disasters as "exceptional", one-off events, scholars in critical disaster studies encourage us to see them as deeply connected to social, political, and economic frameworks. Extreme events reveal and worsen existing inequalities, often reinforcing marginalisation and neoliberal governance patterns. Therefore, this chapter explores alternative responses to extreme events that consider diverse cultural contexts and approaches, from both theoretical and empirical perspectives. These include possibility, hope and care rooted in non-capitalist practices, preparedness, improvisation not just as reactive measures but as political acts, and the broken-world thinking approach.

Keywords Prefigurative politics · Polycrisis · Improvisation · Broken-world thinking · Care and hope · Politics of possibility and proposition

S. Armondi and G. Pessina, *Critical Spatial Thinking and Extreme Events*, PoliMI SpringerBriefs, https://doi.org/10.1007/978-3-032-18875-5_4

4.1 Re-envisioning Extreme Events: Resilience and Recovery

Extreme events are often portrayed as moments of rupture, crisis, and emergency that demand swift responses to restore stability. Strategies for disaster risk reduction encompass a broad range of policies and interventions throughout the entire life-cycle of infrastructure development, including planning, design, construction, maintenance, and decommissioning, as well as the creation, monitoring, and enforcement of legal frameworks. However, from a critical viewpoint, disasters are paradoxically also opportunities—portals—for reconsidering dominant ideas of development, urbanisation, governance, and emergency management (Solnit 2009; Klein 2007). On one hand, they expose the failures of current systems; on the other, they create room for inquiring alternative policies, practices, and imaginings. This chapter examines how extreme events can be contemplated as sites of radical transformation.

Instead of framing disasters as "exceptional" one-off events, scholars in Critical Disaster Studies, discussed in Chap. 2, encourage us to see them as deeply rooted in social, political, and economic systems. Disasters expose and exacerbate existing inequalities, often reinforcing marginalisation and disparities that are amplified by a neoliberal mode of governance (Peck and Tickell 2002). Yet, they also offer opportunities to reorient resilience and recovery, challenging conventional thinking and experimenting with new approaches (Escobar 2018; Gaillard 2018). At their core, disaster risk reduction efforts should aim at fostering systemic social change that prioritises the needs and rights of the poor, marginalised, and oppressed. Numerous documented examples demonstrate how specific disaster risk reduction strategies have successfully prevented major disasters or significantly mitigated their impacts.[1]

The dominant paradigm of disaster preparedness is rooted in risk management frameworks, technical solutions, and forecasting. However, critical disaster studies and urban political ecology suggest that preparedness for extreme events should not only include technical fixes but also reimagine social and spatial structures in a polycrisis scenario, in ways that prevent disasters from perpetuating the same weaknesses and vulnerabilities.

4.2 The Politics of Hope and Care

According to the prominent account of the over-theorisation of capitalism outlined by Gibson-Graham (1996, 2006), scholars such as Fickey (2011) and Cretney (2017) identified the theme of hope: they argued that fostering hope involves recognising

[1] See "Disasters Avoided," https://disastersavoided.com; other key reports on disaster policy are: Intergovernmental Panel on Climate Change Report (IPCC 2022) focused on a number of experiences on how climate-related disasters are unequally distributed, with marginalised communities most affected and OXFAM (2024) on locally led emergency response fund project (from Bangladesh, Puerto Rico, and Uganda).

and strengthening the non-capitalist relations already embedded in everyday life. The concept of hope in social sciences underscores the importance of recognising everyday forms of experimentation and visioning that seek to establish alternative modes of economic, spatial and social organisation (Anderson and Fenton 2008; Head 2016; Pickerill and Chatterton 2006). Within this framework, disruptions to the main socio-economic order, such as those triggered by extreme events, can be understood as "generative moments that harbour the potential to engender transformative possibilities" (Head 2016, p. 166).

Rather than embracing ingenuous optimism (Cameron and Hicks 2014), this perspective reflects a deliberate and sustained commitment to praxis and struggle aimed at materialising alternative socio-economic paradigms beyond the boundaries of capitalism and neoliberalism, and beyond the debate between propositional and oppositional actions (Baptista and Cirolia 2022). Abdoumaliq Simone referred to a specific spatiality: the *surrounds* as concurrently "the spaces, times, and practices within and beyond capture, where experiences of detachment from the predominant forms of accumulation and dispossession enable the specification of dispositions in ways that ward off being apprehended in any definitive sense—where moments of suspension from the traumas and obligations occasioned by individual and collective memory are momentarily possible" (Simone 2022, pp. viii–ix).

As Pieterse echoed in a public lecture at the University of Cape Town (African Centre for Cities 2018), "being both critical and propositional is not a choice, but a necessity". Consequently, southern scholars are compelled to engage in an anticipatory scholarship that not only criticises but also experiments, imagines, and inspires (Mbembe and Nuttall 2004), particularly in the context of southern, decolonial, urban debates.

By valuing these often-invisible economic activities, anti-capitalist hope is rooted in tangible initiatives that embody solidarity and sustainability rather than profit. Likewise, Gibson-Graham et al. (2013) argued that treating capitalist hegemony as totalising is a mistake. Instead of recognising the existing formal economic and market structure, they accentuated the transformative potential of the multitude of minimal socioeconomic practices.

This theoretical alignment inherently stresses prefigurative politics. Prefigurative politics, and the idea of building alternatives through prefiguration, were widely linked to social movements that flourished after the 1960s, drawing on anarchist-inspired principles such as participatory democracy, inclusivity, and direct action (Fians 2022). Local collective actions shape "the seeds of the future in the present" by creating institutions and relationships that promote the values of equality, cooperation, and ecological care they aim at, thereby creating interstitial spaces for change to happen and extend beyond the margins (Wright 2013). Instead of merely opposing the status quo, they actively construct community economies and alternative ways of living. The concept of hope, then, is not naive utopianism oblivious to privileges and power dynamics, nor a warm, fuzzy emotion. Gibson-Graham referred to it as "politics of possibility," rooted in everyday production and exchange practices (Gibson-Graham 2006; Cretney 2017). Through the challenging yet hopeful work of forging alternative socio-economic imaginaries, anti-capitalist actions maintain momentum and affirm that another world is not only possible but already, always, in progress.

This promising imaginary outlines a political project that recognises and actively affirms the magnitude of expanding non-capitalist relations, practices, and values. Such interventions appear through various economic arrangements and the development of practices that resist and operate beyond capitalist logics, along with the use of prefigurative organisational strategies intended to reconstitute and renegotiate dominant societal norms, values, and practices (Pickerill and Chatterton 2006).

While these practices are deeply rooted in tangible realities, the politics of hope does not ignore or neglect experiences of struggle and grief. Instead, it is possible to stay critically attentive to the continual development of spaces and practices that foster the possibility of transformative socio-political change beyond capitalist and neoliberal systems. Hope in this sense is labour—as Freire (1994) noted, it requires active practice. This posture opens a vital framework for rethinking our response to extreme events. Within the realm of environmental justice, prefigurative politics appears in environmental movements like *Extinction Rebellion (XR)* in several countries or *Les Soulèvements de la Terre* in France: not only do they protest against environmental degradation but also advocate for alternative organisational models and communities centred on ecological care and social fairness. More than mere resistance, as Evans (2021) put it, *Extinction Rebellion* movement are gaining strength through their approach to confronting the future through innovative actions and experimentation.

Some scholars recognise colonialism in energy and ecological transitions (Müller 2024). They continue to adopt a techno-managerial approach based on the material "addition and accumulation" of lower-carbon infrastructures within the current energy system. Energy extraction and infrastructure contribute to racist colonial capital and state-led accumulation (Lohmann 2024), upholding systems of colonial domination. The "green economy" imperative exacerbates the problems underlying current socio-ecological crises (Dunlap and Jakobsen 2020). Against this backdrop, it is vital to emphasise the role of care practices in Puig de la Bellacasa's research (2011, 2012, 2017), both in disaster preparedness, planning, and policy responses, by fostering flexibility and anti-fragility in considering spaces and employing unconventional and unexpected (spatial) solutions during crises when plans and projects often fail. Here, the focus is on exploring hopes and possibilities for ecological reparation and care, highlighting the fragile everyday materialities that connect humans and non-humans. According to Papadopoulos et al. (2023), in this context, the notions of care and repair take on a salient role. While recognising the intrinsic fragility of the world, of regions and places, they allow us to highlight another aspect: life may be maintained through subtle, elusive yet "minor acts" and processes of repair, healing, care and reinvention enacted by both humans and non-humans, despite unpredictable—and perhaps unavoidable—experiences of crises and breakdowns (Puig de la Bellacasa 2022). Therefore, framing responses to extreme events as "matters of care" requires a more profound transformation of mainstream emergency responses and the fundamental principles and logics underpinning colonial capitalism and modernity. In this vein, Soto-Hernández (2025) contrasts the colonial capitalist narrative of extraction in the Atacama Desert in Northern Chile with the Lickanantay communities' efforts to shape the Atacama Desert into a plentiful and sacred place, thereby fostering alternative ways of grasping how to care for the territory.

Focusing on the multitude of trivial and minor practices of care and repair diverts attention from large-scale environmental remediation projects or state-led reparation actions, which mean to address ecosystem damages that may have wide-reaching and unpredictable consequences for both local and distant ecological relations. Instead, connecting ecological relations with reparations suggests a shift from top-down, individualised, and abstract interventions to processes of repair that engage with material relations in their complex embeddedness within local contexts.

However, what relevance do they have to the various disciplines that underpin this book? As authors, we believe that all of them are essential for fostering greater sensitivity in education programmes, projects and policies, especially in a world faced with multiple crises—what the anthropologist Anna Lowenhaupt Tsing has called in one of the most acclaimed books on climate change: "the arts of living on a damaged planet" (Tsing et al. 2017).

In 2015, Anna Lowenhaupt Tsing published a remarkable scholarly work titled *The Mushroom at the End of the World: On the Possibility of Life in Capitalist Ruins*. This book depicts a multispecies ethnography that critically challenges the anthropocentric frameworks commonly underlying social science research. Such frameworks often neglect the roles of nonhuman actors in shaping events and outcomes. Tsing's engagement with the global trade of matsutake mushrooms—an industry that gained prominence during the 1980s to meet the demand of Japanese consumers for a culturally significant fungus that had become scarce in Japan—serves as a central example. Through meticulous narrative, the anthropologist relates this economic phenomenon to human cultural practices of foraging, thereby illustrating the entanglements among ecological, economic, and social processes. As a fungus originating from degraded landscapes, matsutake symbolises the potential for ecological renewal and offers a tentative prospect of hope amid prevailing uncertainties. Tsing asserts, "To follow matsutake guides us to possibilities of coexistence with environmental disturbance" (2015, p. 4). This methodological approach, which gives importance to co-constitutive relations between humans and nonhumans—particularly emphasising the agency of nonhumans—serves to expose the misconception of a strict human-nature dualism.

4.3 When Policy and Planning Fail Apart

The concept of resilience has become an ambiguous buzzword in policy and planning actions related to preparing for and responding to extreme events. Resilience is often viewed, in neoliberal economies, as an individual or local responsibility, shifting the burden of adapting to risks onto communities rather than addressing the structural drivers and root causes of disaster vulnerability (Peck and Tickell 2002). Critical scholars have called for resistance-based approaches that introduce a political struggle into disaster preparedness. Embracing resistance in policy means rejecting a return to the pre-disaster status quo when that condition is itself unjust for vulnerable territories. Instead of simply assisting communities to "bounce back better",[2] it questions how they can "bounce forward", by using the disruption and uncertainty of disaster as a chance to push for structural changes that reduce future risks (Manyena et al. 2011). The focus shifts to grassroots collective action, solidarity economies, and empowering marginalised groups as essential to building capacity for change in development policies.

Extreme events upset ordinary planning systems, forcing local communities to experiment with unplanned survival strategies and infrastructures. For example, after the earthquake in Haiti (2010), informal settlements grew as government-led reconstruction efforts failed to satisfy housing needs (Schuller and Maldonado 2016). Similarly, during the COVID-19 pandemic, urban responses saw informal economies and policies—such as for schools and outdoor areas of cafés and restaurants—adapting more quickly than state-led initiatives, with street vendors and community kitchens stepping in to address urgent food and healthcare needs (Armondi et al. 2023).

The framing[3] of urban resilience by key actors has profound implications for the types of policy responses and investment strategies that urban governance structures legitimise, as well as for the distribution of benefits among urban populations (Davoudi et al. 2012). The case of New York City in the aftermath of Hurricane Sandy (2012) illustrates this dynamic particularly well. Fainstein (2018) contrasts two distinct municipal frames for labelling resilience: former Mayor Michael Bloomberg's *A Stronger, More Resilient New York* (2013) and Mayor Bill de Blasio's *One New York: The Plan for a Strong and Just City* (2015). Bloomberg's strategy predominantly accentuated infrastructural resilience, encouraging redevelopment in flood-prone zones and leveraging capital investment from affluent sectors to expand the city's tax base for targeted adaptation projects. In contrast, de Blasio's strategy adopted a broader conceptualization of resilience, incorporating socio-economic dimensions such as income inequality, affordable housing, and access to education—factors widely recognised in the literature as critical to enhancing systemic adaptive capacity and overall human well-being (Eakin et al. 2014). These divergent approaches disclose how differing interpretations of resilience among political leaders can lead to contrasting urban priorities, investment patterns, and ultimately, differential impacts across social groups.

Moreover, in the book Lifehouse (2024), Adam Greenfield critically examines the aftermath of Superstorm Sandy in New York as a cautionary case study of the systemic

[2] This approach, widely adopted in a range of urban policy, organisational and economic studies, is founded on the notion that resilience (the most ubiquitous term in disaster risk management) represents "bouncing back" to a real or felt stable state in recovering from shock and disasters (Manyena et al. 2011; Bondeli and Havenvid 2022). Many studies posit the notion that responses to such external shocks may even make organisations, communities and/or individuals stronger by bouncing "forward" (Manyena et al. 2011) and "back better" (Matyas and Pelling 2015).

[3] Bosomworth (2015) define frame as cultural and political devices that people, institutions, and organizations use to interpret an issue or situation, rank responses, and make investments in those actions.

vulnerabilities in contemporary urban environments. The author stresses how the storm exposed the inherent fragility of infrastructures presumed to be robust, such as power grids, transportation networks, and communication systems, all of which rapidly collapsed under the strain of that extreme event. Greenfield contends that the hyper-centralised and technologically intricate systems characterising contemporary Western cities are ill-equipped to withstand extreme events, precisely because of their scale and opacity. Rather than placing full faith in further technological control or institutional emergency organisations (such as FEMA) or philanthropic agencies (such as the Red Cross), Greenfield emphasises the crucial role of "Occupy Sandy" a self-organised initiative set up in real time by participants of the previous year's Occupy Wall Street resistance and social movement. In *Lifehouse*, Greenfield stresses not only Occupy Sandy but also advocates for a paradigm shift toward decentralised, community-oriented models of resilience. The author first considers the legacy of the Black Panther "survival programs" in the US. The Black Panther Party developed numerous social programs between 1967 and 1982, addressing the immediate and essential needs of the black community. Historians often undervalued these groundbreaking acts, which aimed at improving the conditions of the people, as they were the first to inspire nationwide programs to aid many exploited and subaltern groups. Second, the author considers the role of the "solidarity clinics" experience—a grassroots movement that provided free primary healthcare to people during the 2010–2015 anti-austerity protests in Greece.

Conventional approaches to disaster preparedness remain dominated by Western skills and abstract, technocratic modelling, which privilege expert-driven, top-down solutions and often disregard the knowledge of those most affected. A decolonial critique argues that this paradigm imposes a narrow set of epistemologies, silencing Indigenous communities and local ways of knowing (Gaillard 2018). In practice, the dominance of Global North expertise can lead to culturally inappropriate interventions that are blind to local context (see Chap. 2).

For example, the postcolonial anthropologist Arturo Escobar argues for autonomous, community-led adaptation strategies that challenge dominant colonial, Global North technocratic solutions to development (Escobar 1995) and address underlying power imbalances (Escobar 2018). To prepare for the climate crisis in a mode of resistance thus means not only improving emergency response at the local level, but also mobilising to confront the colonial, racial, and economic structures that render some communities systematically unsafe and uneven.

Excluding local communities' input from early warning systems and infrastructure planning often neglects valuable place-based and Indigenous knowledge about weather, land use, and social organization. This neglect reinforces epistemic hierarchies rooted in colonial legacies, favouring dominant frameworks and marginalising alternative perspectives. Incorporating Indigenous and local knowledge is then more than symbolic; it is decisive for creating fairer and more effective climate adaptation strategies. For example, in the Andean highlands, Indigenous water governance shows how traditional ecological knowledge can address modern socio-environmental issues. Communities here manage water using customary institutions based on reciprocity and conservation (Boelens 2016). Techniques like *amunas*—mountainside infiltration canals—illustrate how ancestral practices of "sowing and harvesting" water can sustainably recharge aquifers and sustain streamflow during dry seasons.

These practices, grounded in a holistic understanding of ecosystems and complex cultural relationships to the land, have enabled Andean farmers to adapt to climate variability for centuries. Integrating such approaches into contemporary adaptation plans can improve water security and distribute resources equitably, all while respecting Indigenous rights and cosmologies.

Indigenous peoples' ecological stewardship provides insights into hazard mitigation beyond water management. Aboriginal fire management (e.g., cultural burning) in Australia shows how traditional practices can reduce disaster risk. For thousands of years before the late 18th-century colonisation, Indigenous Australians used controlled burns to manage landscapes, fostering ecological health and reducing wildfire risks (Bardsley et al. 2019; Singh and Srivastava 2024). This regime of regular, low-intensity burns created resilient landscapes and limited the occurrence of massive wildfires. Colonial authorities suppressed these practices, which contributed to the dangerous fuel build-up that has fuelled today's catastrophic bushfires. As climate change worsens wildfire seasons, there is increasing recognition that integrating Aboriginal fire knowledge with remote sensing and early warning systems can improve ecosystem resilience. Vázquez-Varela et al. (2022) focused specifically on traditional fire knowledge, examining the scholarly literature. Their paper discloses the gap in current research on local fire knowledge among non-Indigenous populations. Consequently, introducing Indigenous-led burning programmes acts as a form of extreme events management that aligns scientific wildfire management with Indigenous stewardship values. However, the disaster is an opportunity, yet another, and always the most powerful one, to manage power unilaterally, because the neoliberal model is a bandwagon with an inertial drive that admits of adjustments but not slowdowns.[4] Listening to native knowledge is not enough if there is no political humus to accommodate their priorities.

4.4 Improvisation as a Crucial Element of Extreme Events Responses

Improvisation in emergency management refers to the ability of individuals and organisations to adapt to unforeseen challenges by developing novel solutions without formal procedures (Kendra and Wachtendorf 2003; Ravazzi 2024). While

[4] On 14 October 2023, Australians voted on a proposal to alter the Australian Constitution to recognise the First Peoples of Australia by establishing an Aboriginal and Torres Strait Islander Voice. The referendum proposed a constitutional amendment to establish a representative body made up of Indigenous Australians to advise the federal parliament on matters affecting the Indigenous community. It was Australia's first referendum in the twenty-first century. The referendum did not pass.

traditional emergency planning emphasises control, structure, and pre-defined roles, disaster events often necessitate rapid decision-making that goes beyond standard protocols. Studies of extreme events, such as September 11, 2001, terrorist attacks (Wachtendorf et al. 2015) and Hurricane Katrina (Rodríguez et al. 2006), illustrate how emergency responders, volunteers, and affected communities frequently depend on improvisation when institutional responses fall short.

According to Quarantelli, in a study that emphasised how planning and management principles should differ for disasters and catastrophes, "Any kind of crisis requires imagination in responding. But the most is required by a catastrophe" (2000, p. 5). According to Glantz and Ramírez (2018), there is likely to be an expected "gap in time" where plans do not matter and improvisation and innovation efforts take over, a *hiatus* in which individuals, groups, and communities no longer wait as ill-fated victims for outside, top-down help. While post-disaster actions rely on rigid bureaucratic structures, disasters often rely on improvisation, adaptation, and creativity. The effectiveness of bottom-up, informal response strategies during disasters presents evidence against the belief that only pre-planned interventions work, however revealing that they surpass top-down governmental responses (Solnit 2009). The unpredictable nature of catastrophes tests standard response systems that depend on fixed plans and bureaucratic procedures. Emergency management requires real-time adaptation and innovation capabilities, which are related to improvisation, but this critical aspect remains frequently unrecognised.

Drawing from Kendra and Wachtendorf's (2003, 2006) seminal work on improvisation in emergency response, we learn how emergency actors navigate uncertainty, how improvisation interacts with institutional constraints, and how critical disaster studies frame adaptive responses to crises. "Improvisation is a significant feature of every disaster … [It] has had something of a checkered history in the emergency management field since its appearance in a disaster response seems to suggest a failure to plan for a particular contingency. Even scholars who have recognised the value of this capacity have tended to subordinate it to planning" (Kendra and Wachtendorf 2006, p. 21).

Improvisation is not just a pragmatic condition or unavoidable practice, but a function of knowledge, resource distribution and systemic inequities (Tierney 2019). Indeed, the ability to improvise is often contingent upon access to expertise, authority and resources, factors that are unevenly distributed across different communities and social groups (Barrios 2017a, b). For marginalised populations, improvisation is about experimentation to survive without state support (Gaillard 2018).

Critical disaster studies promptly point out improvisation as a crucial practice offering an alternative to fixed hierarchical disaster response methods, which often overlook the unpredictable and uneven distribution of crisis effects. Moore (2015) coined the term Capitalocene to replace the notion of Anthropocene, defining the dominant force shaping our epoch through global actors, alongside capitalist investments and significant infrastructural developments involving the exploitation of natural resources and labour. Other scholars have maintained the concept of the Extractive Age (Shapiro and McNeish 2021). This concept reveals the complex nature of an era in which human extraction and utilisation of natural resources contribute to a planetary environmental crisis. These perspectives underscore the importance of improvisation in responding to systemic failures in governance and resource allocation. Scholars argue that extreme events should not be viewed as isolated, exceptional incidents that occur rapidly, but as ongoing processes deeply embedded within socio-economic and political frameworks (Tierney 2019; Barrios 2017a, b; Pelling and Dill 2010). In this context, improvisation is not merely about emergency response but also about empowering communities, activists, and local institutions to stay with the troubles (Haraway 2016), manage uncertainty, mobilise resources, and challenge dominant paradigms of disaster management.

Improvisation in responses to disaster often originates from the ground, led by grassroots activists, marginalised low-class communities, and local actors who are most affected by crises but, at times, receive the least support or a delayed response from official systems. Too often, state-run disaster management prioritises economic recovery over social equity, thereby reinforcing existing inequalities. Consequently, vulnerable communities are left to "improvise" and develop their adaptive solutions (Schuller and Maldonado 2016). These solutions may include mutual aid networks, informal housing for displaced individuals, or the unconventional use of infrastructures and technologies when public intervention falls short. However, there is a risk in how improvisation is regarded. It can be celebrated in ways that shift the burden of survival onto individuals and communities, thereby absolving governments and corporations from responsibility. Neoliberal policies can exploit these grassroots responses to justify a disinvestment in public responsibility (Adams 2013; Klein 2007). Therefore, it is essential to differentiate between a liberatory form of improvisation that empowers communities—helping them reclaim agency and envision alternative futures—and forms that exacerbate precarity and lead to the privatisation of the commons.

To conclude, engaging with improvisation requires a rigorous analysis of power nuances, governance, and the socio-political structures (actors, subjects, sectors, institutions, etc.) that shape both the causes and responses in contemporary societies. Still, improvisation invites us to consider the specificities of each context. Through improvisation, we can appreciate the creative, often subversive ways in which people resist, survive, and transform crisis conditions beyond the constraints of conventional management frameworks. Practically, improvisation helps urban planners and policymakers to continually re-evaluate the categories of actions and relationships to appreciate the diversity and differences of actors.

Institutional Constraints and the Limits of Improvisation

Although improvisation plays a crucial role in disaster response, it is often hindered by slow-moving bureaucracies, rigid systems, and cultures of risk aversion (Dynes 1994). Government agencies and humanitarian organizations tend to fall back on standard procedures, even when those approaches do not fit the unpredictable and complex nature of real crises (Clarke 1999). This sets up a persistent tension: how can institutions remain flexible enough to adapt to the moment, while still maintaining accountability and coordination (Comfort et al. 2010)? At the same time, not all improvisation is considered by the same token. Power imbalances influence which responses are taken seriously and which are overlooked or dismissed. Grassroots efforts—such as the widespread mutual aid networks that emerged during the COVID-19 pandemic (Armondi et al. 2023)—exhibit how communities build resilience through improvisational practices. Yet these contributions are often undervalued in formal disaster planning, despite being central to how people survive and recover. However, despite their effectiveness in crisis conditions, state-led interventions usually dismiss or overshadow these efforts (Simone 2019).

Roux-Dufort and Vidaillet (2003) called attention to the challenges of improvising during a general crisis. They proposed three key considerations: (1) the emergence of improvisation as an unexpected practice depends on crossing critical thresholds, beyond which conditions that initially enable improvisation become obstacles; (2) the involvement of multiple groups with strong professional identities leads individuals to stick to their established roles, thereby limiting adaptive behaviour; and (3) the presence of partial collective interaction and reinforced intra-group dynamics hampers the collective sensemaking processes essential for effective improvisation.

Given the increasing complexity of contemporary extreme events, emergency governance must move beyond rigid structures and embrace adaptive and incremental frameworks that integrate improvisation into formal response mechanisms. Key policy recommendations include:

- Institutionalising flexibility: Cutter et al. (2008) focused on the agency in disaster responses: it should incorporate adaptive learning that allows for real-time adjustments in crisis, allowing coexistence and the sharing of experiences.
- Recognising grassroots innovations: Aldrich (2012) pointed as crucial the acknowledgement of informal and community-led responses; they should be integrated into official emergency planning.
- Addressing power and equity in improvisation: According to Peek et al. (2020), disaster governance organisation should critically examine who can improvise and who is constrained by systemic barriers, ensuring adaptive capacities are equitably distributed.

Improvisation is not just a reaction to unexpected events but functions as a heuristic device as the next section will show. It is a window into broader socio-political dynamics of crisis governance. As crises become more frequent, interconnected and complex, emergency management must embrace critical, adaptable approaches for how communities, first responders and institutions respond to uncertainty by recognising improvisation as a central aspect of disaster response, rather than an exception.

Cases of Improvisation in Top-Down and Bottom-Up Responses

Improvisation in disaster policy responses has become, unexpectedly, an increasingly recognised and essential approach when dealing with complex, unpredictable crises. Traditional top-down disaster management frameworks often depend on uniform protocols and inflexible bureaucratic structures that find it difficult to adapt to the rapidly changing nature of extreme events, especially in areas like climate change, pandemics, and socio-political disruptions. Improvisation, whether applied by governments, humanitarian organisations, or local communities, can improve responses by enabling flexible, adaptable, and context-specific solutions. This section explores key cases where improvisation has played a vital role in shaping disaster policies and responses.

- Hurricane Katrina: Grassroots mutual aid and community-led reconstruction.

Hurricane Katrina (2005) exposed both the failures of government-led disaster response and the potential of grassroots improvisation during crises. The Federal Emergency Management Agency (FEMA) was widely criticised for its delayed and inadequate response, leading to the rise of community-driven initiatives such as the Common Ground Collective, which offered emergency medical care, food, and housing assistance in affected areas (Adams 2013). Local communities and volunteers improvised shelter solutions, organised food distribution networks, and created alternative communication infrastructures when official channels failed. These bottom-up efforts underscored the importance of decentralised, community-led disaster resilience strategies.

- The 2010 Haiti earthquake.

The devastating 2010 earthquake in Haiti led to one of the most significant and substantial humanitarian relief efforts in history, with international and local actors having to improvise in response to logistical bottlenecks and governance failures. Due to the collapse of formal infrastructure, humanitarian organisations resorted to alternative distribution methods, including direct cash transfers to affected populations instead of traditional in-kind aid (Schuller and Maldonado 2016). Additionally, makeshift housing solutions emerged as a mixture of informal settlements, NGO-built shelters, and locally developed structures, reflecting the improvised strategies used by displaced populations when formal reconstruction efforts stalled (Farmer 2011). These examples foreground how improvisation in response often arises out of necessity, mainly when formal structures are weak or overwhelmed.

- Japan's 2011 triple disaster.

The 2011 Tōhoku earthquake, tsunami, and Fukushima nuclear disaster prompted Japan to revise its national energy and disaster recovery policies quickly. The nuclear meltdown at Fukushima Daiichi required urgent improvisation in radiation containment measures, including the ad hoc use of water injections to cool reactors and the construction of makeshift seawalls (Funabashi and Kitazawa 2012). In the aftermath, Japan reformed its energy policy, increasing investments in renewable energy as an improvised response to the sudden loss of nuclear power capacity. The government also adapted housing and infrastructure recovery plans based on local community needs rather than pre-existing national frameworks, demonstrating the value of policy improvisation in post-disaster governance (Aldrich 2019).

- Puerto Rico's 2017 Hurricane María.

The contrast between individual resilience and resistance became clear in Puerto Rico after Hurricane María. Despite insufficient government aid, Puerto Ricans organised spontaneous mutual aid centres, community kitchens, and volunteer brigades to rebuild infrastructure and care for their neighbours (Bonilla and LeBrón 2019). These grassroots initiatives went beyond the concept of resilience-as-survival. They demonstrated resistance by creating new social relations of solidarity and directly challenging the neglect rooted in Puerto Rico's colonial political status and austerity policies (Pinchin 2019). Mutual aid networks supplied food, water, and electricity in affected areas and acted as platforms for community members to voice their dissent and demand a fair recovery. In this way, post-María Puerto Rico showed how disaster preparedness rooted in resistance can promote deeper social change, inspiring community autonomy, politicising the recovery process, and pressuring authorities for long-term, equitable reconstruction (Pinchin 2019).

In the account of ethnographic fieldwork, Garriga-López (2019) contended that the ongoing condition of US colonialism fundamentally limits the potential for meaningful disaster preparedness and recovery in Puerto Rico: "Puerto Rico has become a highly active extractive zone on the periphery of US empire" (p. 174). Although the future path of the archipelago remains uncertain, specific cultural dimensions of disaster recovery articulate a decolonial imaginary by promoting solidaristic practices among communities. In this context, although post-hurricane recovery efforts were hampered by systemic dysfunction, militarised intervention, and the reinforcement of imperial authority through political manipulation, this disaster also reveals possibilities for decolonial praxis. They act as an essential "litmus test" for applying and examining decolonising frameworks and narratives as part of the broader reconstruction process. Bonilla (2020) demonstrated, through the lens of the "coloniality of disaster," how catastrophic events, such as hurricanes, fires, and earthquakes, can either reinforce existing colonial structures or open opportunities to rethink sovereignty and community resilience. Local grassroots projects, such as Casa Pueblo's solar energy projects alongside *Comedores Sociales*[5] exemplify how communities can adopt decolonial recovery strategies centred on self-sufficiency and oppose externally imposed top-down initiatives that perpetuate dependencies.

- Temporariness, voids, and improvisation in urban contexts during the COVID-19 pandemic.

The unpreparedness for the COVID-19 pandemic underscored the crucial role of urban governance in improvising. Cities worldwide faced exceptional challenges that demanded immediate and flexible responses. Through their investigation, Caramaschi and Coppola (2023) demonstrate how urban actors utilised temporary structures through spontaneous tactics and varied uses to adapt to pandemic-related uncertainties. The research findings enhance critical disaster studies by illustrating how urban spaces interact dynamically with adaptive practices during unforeseen crises. The study reveals how vacant spaces are vital in developing temporary solutions amid uncertainty. Lockdowns and social distancing measures brought to the forefront the

[5] Both organizations emerged long before Hurricane Maria.

importance of leftover and empty urban spaces, which facilitated transient interventions and unexpected uses. The adaptive practices involved the utilisation of vacant spaces for public health measures, along with the support of various urban actors' operations and the introduction of new, flexible, and temporary options. The study by Caramaschi and Coppola offers valuable insights into the role of improvisation in urban governance during crises. Their research suggests that integrating temporariness and flexibility into policy and planning tools can improve urban preparedness by enabling the population (and their "crafty" solutions) to adapt swiftly to emerging challenges.

- Heatwaves and Climate Adaptation in Southern Europe.

Considering the escalating heatwaves across Southern Europe, which are attributed to climate change, urban areas have been compelled to develop novel adaptation strategies aimed at safeguarding vulnerable populations, particularly the elderly and economically disadvantaged residents. Barcelona's implementation of climate shelters exemplifies this approach: in the absence of comprehensive national policy measures, the city has established a network of "climate shelters" within public buildings, such as schools, to serve residents during episodes of extreme heat. The Barcelona Municipality has exemplified leadership by acknowledging the repercussions of climate change and committing to "unexpected" measures. These initiatives encompass planning to mitigate vulnerabilities, leverage existing strengths, enhance resilience, and attenuate the impacts of climate change. The project, managed by the City of Barcelona and supported by the European Commission's Urban Innovative Actions programme, was operational from 2019 to 2022.[6] Although some European nations have advanced environmental policies, environmental degradation persists at a slow pace due to regulatory loopholes, corporate influence, and political stagnation within the European Union. Consequently, improvisational tactics can become crucial to address these regulatory deficiencies through grassroots activism, municipal initiatives, or experimental policy development.

The ongoing requirement for improvisation highlights deeper, fundamental systemic breakdowns. As the previous vignettes illustrated, various improvisation strategies have been partially institutionalised through policy and planning measures. However, others have demonstrated limited government intervention in addressing environmental protection needs. Within a neoliberal framework, the gradual decline in public support means that governments are unable to establish sufficient long-term solutions. Thus, local communities become responsible for environmental protection, but this can lead to increased social inequalities.

Policymakers must recognise improvisation as a vital part of adaptive environmental governance alongside principles of justice, rather than merely seeing it as an emergency patch. By incorporating approaches that permit improvisation into formal policy frameworks and maintaining accountability and fairness, urban and regional policies can more effectively address the ongoing and evolving threats of slow environmental violence.

[6] https://portico.urban-initiative.eu/urban-stories/uia/barcelona-city-climate-action.

Can "Improvisation Expertise" Be Learnt?

These cases, as mentioned earlier, call attention to the extent of improvisation in disaster policy responses across various contexts. Whether in adaptive public health measures, grassroots-led mutual aid efforts, or flexible governance adjustments, improvisation is vital for navigating crises and disasters when formal institutional frameworks are inadequate. However, the role of improvisation is not without its jeopardy. For example, it can be exploited, marketed or co-opted by neoliberal emergency models that shift responsibility onto individuals and communities, and it may also reinforce inequalities if access to resources and adaptive capacities are unevenly distributed. Therefore, improvisation can offer a "liberatory" and decolonial potential, addressing structural limitations and ensuring that adaptive strategies lead to more just, resilient response systems. What a colonial perspective regards as chaotic improvisation is merely a way to "know otherwise", to engage more deeply with different peoples and cultures, willing to understand and to "accept unexpected knowledge" (da Silva et al. 2021, p. 7). These dimensions explain why, with a decolonial lens, improvisation involves "unlearning in order to learn" (Tlostanova and Mignolo 2012).

Lindblom and Cohen (1979) conceptualised *usable knowledge* as that which originates outside the formal structures of professional inquiry, knowledge derived not from technical methodologies or rigorous scientific testing, but rather from common sense, casual empiricism, and reflective speculation and analysis. According to Lindblom and Cohen, social scientists do not always fully appreciate that social science inquiry is only one of several ways to disentangle wicked social problems. They are also often engaged in a mistaken search for authoritativeness, denying that their contribution is constitutively incomplete and partial. Glantz and Ramírez (2018, p. 13) echoed the theme in the field of extreme events by proposing a definition of usable knowledge as "a type of vernacular knowledge, which is unplanned and hands-on, that emerges in response to a quick onset event". Consequently, there are "learning flickers", arising from the tenacity of usable knowledge that governments, communities, and individuals can share and appreciate to mitigate extreme events.

4.5 Moving Beyond Building Back Better: The Broken-World Thinking

The broken-world approach is developed in critical infrastructure and technology studies, which challenge conventional patterns of stability, efficiency, and technological progress. It was first introduced by Jackson (2014) in his work on repair cultures in these fields, where the author contested the hype of innovation and disruption and emphasised the constructive role of repair, maintenance, and improvisation during crises, disruptions, and breakdowns. Instead of viewing disasters as events to recover from or to build back better without tipping points, "broken-world thinking" embraces repair, maintenance, and adaptation as continuous, everyday processes.

The authors help us better understand the "real limits and fragility" of contemporary territories.

Modern and Western environmental movements have often been co-opted by mainstream sustainability (Wahl 2016) and the "green growth" (Kallis and Hickel 2020) refrain. Through the approach of broken planet, Jackson, along with scholars from further research fields—particularly Tsing et al. (2017)—recognises that we live amidst ongoing crises, decay, and incomplete solutions. This perspective shifts the focus from idealised notions of resilience and sustainability, highlighting the importance of repair, adaptation, and improvisation practices that enable people to live within and navigate broken systems (technological, infrastructural, environmental, or social). The core assumptions of broken-world thinking are as follows:

1. Breakdowns and failures are ordinary, not unusual. Outdated and standard views on governance, disaster management, and technological progress assume that failures are rare and should be avoided or repaired. Conversely, Broken-world thinking suggests that failures are constant and unavoidable: infrastructure deteriorates, economies collapse, and environmental systems degrade over time.
2. Maintenance and repair are just as crucial as innovation. Much of advanced policy and planning nurtures innovation, such as developing new systems, technologies, and solutions. However, "infrastructure mediates everyday urban life" (Graham and McFarlane 2015, p. 8), and most daily activities and disaster responses involve maintaining, patching, and adapting existing structures rather than creating entirely new ones. For example, many communities rely on repairing, crafting and adapting existing infrastructure to survive crises rather than building entirely new infrastructure.
3. Improvisation and adaptation are crucial. In a "broken territory", people must make do with what they have, often using informal, temporary, or unplanned solutions. This is especially true in disaster situations, where formal policy responses are frequently ineffective or arrive too late, compelling communities to improvise their own survival strategies.
4. Planning for chronic crises or permacrisis, not just emergency responses. Traditional disaster planning assumes a return to normal after a crisis. The Broken-World Thinking framework recognises that many people, especially in marginalised communities, never experience a stable "normal" and instead live in a permanent state of crisis. For example, climate change causes ongoing, slow disasters, and communities must adapt repetitively rather than wait for one-time recovery.

4.6 Implications for Policy and Planning

Several regions face broken infrastructure systems and services which fail every day. Communities take action to resolve problems by leveraging local expertise and available resources, rather than waiting for government or corporate interventions. Neighbourhoods in Lagos, Nairobi, and Dhaka depend on unofficial electricity supplies and self-created water systems, along with improvised transit networks, due to governmental neglect. According to Simone and Pieterse (2018) cities in Asia and Africa are a kind of "paradoxes of the urban": they serve as thick arenas for a range of countervailing tendencies, sites where contraction and expansion, ambiguity and intelligibility, fragmentation and consolidation, physical embodiment and digital abstraction, as well as movement and stasis, are not only heightened but often converge to the point of becoming indistinguishable. "So much of urban Asia and Africa seems caught in a catch-22[7]" (Simone and Pieterse 2018, p. 25), but these systems may be unstable and inefficient. However, they reflect real-world adaptation strategies when formal systems fail.

In response to geo-shock and geo-trauma, Pain (2021) and Ortiz and Gómez Córdoba (2023) introduced the concept of "territorial healing": a theoretical and methodological framework for studying and guiding territorial actions in communities affected by violence-induced trauma, which can also serve as inspiration for dealing with extreme events in general. Drawing on contemporary debates about decolonial actions and reparative planning, those scholars articulated a set of conceptual foundations and strategies to address the territorial wounds provoked by many forms of implicit and explicit violence, including colonialism, structural oppression, "natural" disasters, armed conflict, and racialised violence. The starting point is territorial "wounds": "Urban settings are weaved not only by the visible wounds of the ruins but also by the urban fears inscribed in the invisible borders, imposed rules of coexistence, the everyday life of a guarded territory or the elimination of community spaces of gathering" (Ortiz and Gómez Córdoba 2023, p. 3). The authors argue that territorial healing demands a holistic approach that interweaves the mapping of "body-territory-earth" relations (Cabnal 2019; Zaragocin and Caretta 2020), the activation of collective memory, and the nurturing of spatial imagination through countermapping. Such a research path enables the redefinition of conflict through therapeutic dialogue and the co-creation of reparative infrastructures rooted in the lived experiences and aspirations of affected communities.

Broken-World Thinking spurs policymakers, engineers, architects, designers and planners to rethink resilience and recovery in several ways:

- Accepting that some systems cannot be restored, emergency planning and policies should focus on continuous adaptation rather than one-time recovery efforts. Instead of repeatedly rebuilding the same infrastructure in areas prone to chronic flooding, policies should invest in adaptive, flexible solutions such as floating homes or relocations. Architecture, planning, and design are central to "caring for the habitat," its inhabitation, and continued livability (Fitz et al. 2019).
- Integrating community repair practices into policy. Local repair networks operate informally, yet they deserve to be considered fundamental elements of policy development rather than backup solutions. Policy developers need to partner with local repair groups and informal engineers who keep systems operational on a day-to-day basis.

[7] A problematic situation for which the only solution is denied by a circumstance inherent in the problem or by a rule.

- Designing infrastructure for maintenance, not just new construction, and recalibrating urban planning practices. This means adding value to repair and maintenance, restoring interstitial spaces, rather than promoting large-scale new project development. Although governments make significant investments in major projects, they often fail to plan for their maintenance after completion. The broken-world thinking approach advocates constructing "flexible" infrastructures that support adaptation to changing conditions.
- Supporting decentralised disaster response and governance. Post-disaster futures should be shaped by affected communities rather than imposed by external actors. Traditional centralised disaster response systems struggle to deliver aid to communities both swiftly and fairly because bureaucratic processes cause delays and lead to inequitable distribution. Policies should promote decentralised emergency response systems that enable community members to organise and develop practical solutions independently.
- Recognising that brokenness is political. Recognising the inequalities in those who must repair and who can afford to ignore a breakdown. Those who experience breakdown—and who get access to repair—are influenced by political and economic inequalities. Emergency planning must address structural issues such as environmental racism, economic precarity state neglect, ensuring policies support vulnerable communities rather than reinforce existing disparities.
- Embracing imperfection and improvisation: If all crisis responses involve working with limited resources, broken systems, chronic disruption, and uncertainty (Simone 2013), policies should be flexible and allow for local, adaptive problem-solving rather than rigid, top-down responses.[8] Instead of treating disasters as exceptional events, this approach integrates them into long-term governance, recognising that many communities, especially marginalised ones, already live in a state of enduring crisis due to environmental degradation, economic precarity, and infrastructure decay (Howe et al. 2016). Consequently, rather than focusing solely on reconstruction after disaster, policies should select ongoing maintenance, acknowledging that infrastructure and social systems are in a constant state of repair (Mattern 2018).

We can view Broken-World Thinking as a "minor" approach (see Chap. 1) because it does not present a utopian vision of stability. Instead, this approach recognises that we live in a world where systems fail, crises endure, and repair is an ongoing process. However, by learning to work with brokenness rather than against it, societies can become more adaptable, just, and resilient in times of uncertainty.

Public policy and planning have usually relied on assumptions of stability, efficiency and the ability to restore systems to a pre-disaster state. However, this approach is being challenged as disasters increasingly become chronic, interconnected and

[8] Simone (2019) examines how residents in urban areas, particularly in the Global South, navigate and adapt to challenging environments through improvisation and adaptability. He portrays urban districts as sites of tireless transformations, highlighting the rhythms that mediate between residents' needs to remain inconspicuous and their aspirations to create unique niches of exception.

driven by systemic failures, particularly in areas such as climate change, environmental degradation and slow violence. The broken-world approach offers an alternative that recognises the inevitability of breakdowns, improvisation and the necessity to work with incomplete, deteriorating or failing systems. In disaster policy and planning, this helps shift the focus from constantly striving for perfect innovation to embracing ongoing repair, maintenance and adaptive transformation in response to new crises. This perspective thus aids in moving disaster policy from top-down recovery to continuous adaptation, making it especially relevant in the era of climate change, where disasters are no longer extraordinary disruptions but part of an ongoing and worsening crisis, as exemplified in the following cases.

- Flint water crisis (2014–present): permanent crisis and community-led repair in US.

The Flint water crisis, a gradual disaster caused by lead contamination and exposure, illustrates how Broken-World Thinking applies to environmental and infrastructural failures. Despite persistent public health threats, the crisis remains unresolved, prompting communities to create their adaptive solutions, such as DIY water testing and filtration. As government agencies minimised contamination risks, residents and independent scientists (like those from Virginia Tech) tested and publicised lead levels in water, raising awareness about lead exposure harm, especially for children (Clark 2017). Mutual aid networks formed to distribute bottled water and install home filtration systems, operating alongside the polluted infrastructure rather than relying on it. Flint's case highlights how, amid slow and continual environmental damage, crisis becomes a perpetual condition. Urban policy cannot depend on quick fixes; it must enable ongoing repair and adaptation. Flint's experience displays that failure and crisis are normalised under slow environmental violence, requiring continuous intervention and adaptive strategies instead of single solutions.

- Climate change and the managed retreat movement in European countries.

As climate change accelerates coastal erosion and flooding in recent years, many regions in Europe are shifting from traditional flood defences to managed retreat, an approach that recognises the inevitability of environmental transformation rather than attempting to resist it. The Netherlands' "Room for the River" project: Instead of reinforcing dikes indefinitely, Dutch policymakers have embraced controlled flooding and the relocation of communities to enable rivers to expand naturally (Klijn et al. 2018). This case reflects a growing recognition that the established assumption of resilience—simply returning to normal after a crisis—no longer aligns with the realities many communities face. As environmental and infrastructure systems continue to break down or shift unpredictably, disaster response must also adapt. To address this, policy needs to move beyond short-term fixes. Planners and policymakers must acknowledge the efforts already underway at the community and local levels, where people are repairing, adapting, and discovering new ways to live in uncertain conditions. This involves accepting that some systems may not be restored and instead focusing on long-term adaptation and structural change. Disasters are no longer isolated events; they form part of an ongoing condition. Policy must evolve accordingly.

- Caring for objects affected by disaster: An ethnography of a recycling centre in Fukushima Prefecture in Japan.

In the small town of Namie (Fukushima Prefecture), evacuated during the triple disaster of 11 March 2011, community ties have been slow to reconstitute, even as new residents begin to settle. Exposure to "low doses" of radiation has profoundly altered people's relationship with their environment. In this broken territory (Wynn Kirby 2023), one of the central challenges is to take stock of the social bonds damaged by the disaster and to consider how they might be repaired. The research of the anthropologist Jeanne Bouillet begins with the singular trajectory of Izumi-san, who returned to Namie after 11 March (https://www.fmsh.fr/en/projects/caring-objects-affected-disaster). In her recycling centre, she collects objects left behind by those who were evacuated following the nuclear accident—items abandoned for nearly a decade. In this damaged world, the act of recovery, which brings objects back into the social realm, may be seen as a form of care for both things and people.

4.7 Extreme Events as a Polycrisis Symptom

We have come to understand that disasters no longer occur in isolation. Climate breakdown, economic instability, geopolitical conflicts, and social inequalities intersect to create a polycrisis, also known as a permacrisis, when multiple crises co-occur and reinforce one another, amplifying the impacts. In the next section, we explore the nexus between disasters and polycrisis, analysing how various risks create cascading failures and how response and recovery efforts must adapt to this reality.

Historically, disaster research focused on single-event hazards such as earthquakes, floods, and wildfires, viewing them as isolated shocks. However, as discussed in Chap. 2, the Urban Political Ecology perspective and Critical Disaster Studies maintain that disasters are complex and systemic phenomena, linked to broader patterns of urbanisation, governance, and social inequality (Klein 2007). In this context, extreme events are not unusual anomalies but rather indicators, or "symptoms," of underlying systemic vulnerabilities.[9] According to Tooze (2021), a polycrisis occurs when multiple crises overlap, mutually amplifying each other's effects and creating unpredictable interactions. The key difference between multi-crisis scenarios and polycrisis situations is that the latter involve deep entanglements that render standard responses ineffective (Dalby 2022).

Key features that explain the extent of a polycrisis are the following:

---

[9] The 2022 flood in Pakistan was both a humanitarian catastrophe and a significant challenge to the nation's debt issues. The disaster occurred during a severe economic crisis, forcing dependency on IMF bailouts. Furthermore, a multiple crisis interaction occurred: the deterioration of food security led to political turmoil and widespread demonstrations. The destruction of infrastructure created delays in the recovery process, leading to a persistent economic reliance.

1. The interaction between climate disasters, economic collapse and social unrest creates a complex network of risks.
2. Small interferences and glitches can set off extensive cascading failures throughout a system, as demonstrated by supply chain disruptions following extreme events.
3. The concurrent or quick succession of multiple crises lessens the time available for recovery efforts.
4. Political inertia occurs when governments face difficulty setting priorities, resulting in denial, slow or insufficient responses.

Additionally, since the concept of polycrisis has become a buzzword in global economic and political debates of the Global North, recent thought-provoking studies have argued that urban and territorial scholars need to rethink the idea of polycrisis, particularly through a decolonial perspective. According to Ruwanpura et al. (2025), colonial legacies continue to guide policy responses to polycrisis: "countries in the global South currently face the double burden of servicing economic debt and meeting the costs of climate change" (Ruwanpura et al. 2025, p. 3).

When Sultana (2022, p. 1) maintains "The extremely uneven and inequitable impacts of climate change mean that differently located people experience, respond to, and cope with the climate crisis and related vulnerabilities in radically different ways," she draws our attention to the possibility of examining the varied responses and vulnerabilities linked to climate change. As such, the concept of polycrisis remains theoretically incomplete until it is reviewed through a decolonial lens, thereby there is a pressing need to develop a new conceptual framework with the integration of diverse experiences, knowledge systems and subaltern interpretative frameworks into the polycrisis.

4.8 Conclusion

To conclude, in the contemporary era, disasters are not unique events, but rather manifestations of interconnected systemic failures. The unfolding of compound crises demonstrates how cascading disruptions transcend spatial and temporal boundaries, intersecting with other crisis phenomena such as natural hazards, climate change, financial instability and socio-political fragmentation (Keck and Sakdapolrak 2013; Biermann et al. 2022). Standard emergency governance models, rooted in siloed and reactive responses, are no longer adequate to address the complexities of polycrisis. Instead, we must adopt an integrated approach to reconfiguring responses, and here the role and interpretation of infrastructures are crucial.

The interplay between infrastructure failures and broader crises reveals how vulnerabilities are embedded within critical infrastructure systems. Infrastructure disruptions in energy grids, water systems or transportation networks do not occur in isolation but trigger cascading failures across sectors and geographies (Pescaroli and Alexander 2018). However, infrastructures are a set of socio-material processes constantly made and remade. According to Graham and McFarlane (2015), a focus on infrastructure "lives" is decisive to an understanding of how infrastructures do not simply exist but "occur"; their role in crises exacerbates pre-existing inequalities, disproportionately affecting marginalised communities that have the least capacity to adapt (Pelling and Dill 2010). Moreover, the temporal dimension of such crises challenges conventional disaster management, as disruptions unfold in non-linear, protracted and often unpredictable ways.

Governing multiple crises and their spatial impacts requires managing fragmented institutional settings and mindsets where a mixture of actors—state agencies, private companies, civil society, and international organisations—hold conflicting interests and operate under epistemic uncertainty (Boin et al. 2020).

The nature of collaboration among actors working towards standard recovery actions—namely, who is involved, how, and why they are involved—is a vital aspect of developing effective responses. In practice, these actors frequently encounter recurring limitations due to power imbalances among stakeholders, contested knowledge systems and governance structures. These factors can reinforce institutional stability rather than promote adaptability. Developing resilience-enhancing strategies requires a shift from centralised technocratic methods to decentralised frameworks that incorporate local understanding and encourage participatory decision-making processes (Meerow and Newell 2019).

This chapter emphasised that merely highlighting the shortcomings of current governance structures is no longer enough. Policies and planning tools to come must urgently rethink political and institutional systems to foster opportunities for transformative change. Governance needs to shift from a reactive approach—dealing with extreme events after they materialise—to proactive systemic strategies that uphold principles of equity and sustainability, while moreover developing long-term adaptive and caring capacities (Bulkeley et al. 2022).

These provocations raise urgent questions for the work to come: How can disaster preparedness frameworks effectively incorporate grassroots and community-led approaches? What does a decolonial and socially just recovery process require? In what ways can urban planning and engineering disciplines contribute to post-disaster imagination? For example, what would it mean to design cities that accommodate instead of resist floods? What if disaster response focused on territorial healing rather than militarised management?

Such questions accentuate the call for collective reflection on how we perceive and respond to extreme events. From this perspective, several planning and policy implications arise. First, strengthening multi-scalar governance mechanisms is essential to facilitate coordination across levels of government and among diverse stakeholders (Duit et al. 2010). Second, investment in infrastructural redundancy and flexible systems can help mitigate cascading failures and enhance adaptive flexibility (Linkov et al. 2014). Third, embedding community-led, place-based adaptation efforts is critical to addressing the structural inequities that often underlie mainstream resilience policy and planning (Matin et al. 2018).

Isolated disasters no longer define our fractured planet. The presence of interconnected crises and reliance on multiple habitats requires a shift away from fragmented solutions towards essential systemic transformation. Improving infrastructure resilience today calls for governance reform that actively addresses the complexities and dependencies in modern risk landscapes, rather than merely focusing on recovery efforts. This change introduces a new policy and planning approach that goes beyond stability to adopt adaptive, incremental actions, recognising territorial fragmentation, opportunities, and redistributing spatial justice, while emphasising their role as "agents" of inclusivity, repair, healing and care.

The concept of extreme events calls for a more radical framing, one that extends beyond risk management to examine systemic spatial injustices and open opportunities for reimagining more pluralistic futures. Subtler, often unseen, yet persistent opportunities for intervention arise in many contexts. This chapter aims at addressing this gap. Improvisation happens not only as a reactive measure but also as a political act, providing informal and community-led strategies that policy and planning frameworks need to recognise and support. Recovery, too, should be reconsidered: instead of restoring the status quo, extreme events can act as critical turning points—portals—through which alternative futures might be envisioned and fulfilled.

References

Adams V (2013) Markets of sorrow, labors of faith: New Orleans in the wake of Katrina. Duke University Press, Durham

African Centre for Cities (2018) ACC international urban conference, 1 Feb 2018. Available from https://www.youtube.com/watch?v=sCyk0yHU5FU

Aldrich DP (2012) Building resilience: social capital in post-disaster recovery. University of Chicago Press, Chicago

Aldrich DP (2019) Black wave: how networks and governance shaped Japan's 3/11 disasters. University of Chicago Press, Chicago

Anderson B, Fenton J (2008) Editorial introduction: spaces of hope. Space Cult 11(2):76–80

Armondi S, Balducci A, Bovo M, Galimberti B (2023) Cities learning from a pandemic toward a preparedness. Routledge, London and New York

Baptista I, Cirolia LR (2022) From problematisation to propositionality: advancing southern urban infrastructure debates. Trans Inst Br Geogr 47:927–939. https://doi.org/10.1111/tran.12537

Bardsley DK, Prowse TAA, Siegfriedt C (2019) Seeking knowledge of traditional Indigenous burning practices to inform regional bushfire management. Local Environ 24(8):727–745. https://doi.org/10.1080/13549839.2019.1640667

Barrios R (2017a) Beyond the disaster exceptionalism: social reproduction and the politics of everyday life in post-Katrina New Orleans. Am Anthropol 119(3):558–570

Barrios RE (2017b) What does catastrophe reveal for whom? The anthropology of crises and disasters at the onset of the Anthropocene. Annu Rev Anthropol 46:151–166

Biermann F, Hickmann T, Sénit CA, Beisheim M, Bernstein S, Chasek P, Viehoff V (2022) Scientific assessments in a polycentric world: a framework for global environmental governance. Glob Environ Change 72:102447

Boelens R (2016) Water, power and identity: the cultural politics of water in the Andes. Routledge, London and New York

Boin A, Ekengren M, Rhinard M (2020) Hiding in plain sight: conceptualizing the creeping crisis. Risk Hazards Crisis Public Policy 11(2):116–138
Bondeli JV, Havenvid MI (2022) Bouncing back in turbulent business environments: exploring resilience in business networks. Ind Mark Manage 107:383–395
Bonilla Y (2020) The coloniality of disaster: race, empire, and the temporal logics of emergency in Puerto Rico, USA. Polit Geogr 78. https://doi.org/10.1016/j.polgeo.2020.102181
Bonilla Y, LeBrón M (2019) Aftershocks of disaster: Puerto Rico before and after the storm. Haymarket Books, Chicago
Bosomworth K (2015) Climate change adaptation in public policy: frames, fire management, and frame reflection. Environ Plan C Gov Policy 33:1450–1466. https://doi.org/10.1177/0263774X15614138
Bulkeley H, Edwards GAS, Fuller S (2022) Governing climate resilience: governing through governance dilemmas. Environ Polit 31(5):805–827
Cabnal L (2019) El relato de las violencias desde mi territorio cuerpo-tierra. In: Leyva Solano X, Icaza S (eds) Tiempos de muerte: cuerpos, rebeldias y resistencias. CLACSO Tomo IV, pp 113–127
Cameron J, Hicks J (2014) Performative Research for a Climate Politics of Hope: Rethinking Geographic Scale "Impact" Scale and Markets Antipode 46(1) 53–71 https://doi.org/10.1111/anti.12035
Caramaschi S, Coppola A (2023) The role of temporariness, emptiness and improvisation in uncertain times: evidence from urban responses to COVID-19. City Territ Archit 10(1):1–15. https://doi.org/10.1186/s40410-024-00245-5
Clark A (2017) The poisoned city: Flint's water and the American urban tragedy. Metropolitan Books, New York
Clarke L (1999) Mission improbable: using fantasy documents to tame disaster. University of Chicago Press, Chicago
Comfort LK, Boin A, Demchak CC (eds) (2010) Designing resilience: preparing for extreme events. University of Pittsburgh Press, Pittsburgh
Cretney RM (2017) Towards a critical geography of disaster recovery politics: perspectives on crisis and hope Abstract Geography Compass 11(1) https://doi.org/10.1111/gec3.12302
Cutter SL, Barnes L, Berry M, Burton C, Evans E, Tate E, Webb J (2008) A place-based model for understanding community resilience to natural disasters. Glob Environ Change 18(4):598–606
da Silva CM, Sauerbronn FF, Thiollent M (2021) Decolonial studies, non-extractive methods, and participatory action research in accounting. J Contemp Adm 26(4):e210014. https://doi.org/10.1590/1982-7849rac2022210014.en
Dalby S (2022) Rethinking environmental security in an era of polycrisis. Cambridge University Press, Cambridge
Davoudi S et al (2012) Resilience: a bridging concept or a dead end? Plan Theory Pract 13(2):299–333
Duit A, Galaz V, Eckerberg K, Ebbesson J (2010) Governance, complexity, and resilience. Glob Environ Change 20(3):363–368
Dunlap A, Jakobsen J (2020) The violent technologies of extraction: political ecology, critical agrarian studies and the capitalist worldeater. Springer International Publishing, Cham. https://doi.org/10.1007/978-3-030-26852-7
Dynes RR (1994) Situational altruism: toward an explanation of pathologies in disaster assistance. Am Sociol Rev 59(4):205–215
Eakin HC, Lemos MC, Nelson DR (2014) Differentiating capacities as a means to sustainable climate change adaptation Global Environmental Change 271–8 https://doi.org/10.1016/j.gloenvcha.2014.04.013
Escobar A (1995) Encountering development: the making and unmaking of the third world. Princeton University Press, Princeton
Escobar A (2018) The making of social movements in Latin America: identity, strategy, and democracy. Routledge, London and New York

Evans H (2021) Beyond resistance: the role of prefiguration in social movements addressing the climate crisis. Bath papers in international development and wellbeing, 66. University of Bath, Centre for Development Studies (CDS), Bath
Fainstein SS (2018) Resilience and justice: planning for New York City. Urban Geogr 39(8):1–8. https://doi.org/10.1080/02723638.2018.1448571
Farmer P (2011) Haiti after the earthquake. Public Affairs, New York
Fians G (2022) Prefigurative politics. In: Stein F (ed) The Cambridge encyclopedia of anthropology. https://doi.org/10.29164/22prefigpolitics
Fickey A (2011) "The focus has to be on helping people make a living": exploring diverse economies and alternative economic spaces: diverse economies and alternative economic spaces. Geogr Compass 5(5):237–248
Fitz A, Krasny E, Wien A (eds) (2019) Critical care: architecture and urbanism for a broken planet. MIT Press, Cambridge, MA
Freire P (1994) Pedagogy of hope: reliving pedagogy of the oppressed. Continuum, New York
Funabashi Y, Kitazawa K (2012) Fukushima in review: a complex disaster, a disastrous response. Bull At Sci 68(2):9–21
Gaillard JC (2018) Disaster studies inside out. Disasters 42(1):S8–S17
Garriga-López A (2019) Puerto Rico: the future in question. Shima Int J Res Island Cult 13. https://doi.org/10.21463/shima.13.2.13
Gibson-Graham JK (1996) The end of capitalism (as we knew it): a feminist critique of political economy. Blackwell Publishers
Gibson-Graham JK (2006) A postcapitalist politics. University of Minnesota Press, Minneapolis
Gibson-Graham JK, Cameron J, Healy S (2013) Take back the economy: an ethical guide for transforming our communities. University of Minnesota Press, Minneapolis
Glantz MH, Ramírez IJ (2018) Improvisation in the time of disaster. Environ Sci Policy Sustain Dev 60(5):4–17. https://doi.org/10.1080/00139157.2018.1495496
Graham S, McFarlane C (eds) (2015) Infrastructural lives urban infrastructure in context. Routledge, London and New York
Greenfield A (2024) Lifehouse: taking care of ourselves in a world on fire. Verso, London and New York
Haraway DJ (2016) Staying with the trouble: making kin in the Chthulucene. Duke University Press, Durham
Head L (2016) Hope and grief in the Anthropocene. Routledge, London and New York
Howe C et al (2016) Anthropology and the politics of repair. Routledge, London and New York
IPCC (2022) Climate change 2022: impacts, adaptation and vulnerability. Report. https://www.ipcc.ch/report/ar6/wg2/
Jackson SJ (2014) Rethinking repair. Media Stud 8(1):221–239
Kallis G, Hickel J (2020) Is green growth possible? New Polit Econ 25(4):469–486
Keck M, Sakdapolrak P (2013) What is social resilience? Lessons learned and ways forward. Erdkunde 67(1):5–19
Kendra JM, Wachtendorf T (2003) Elements of resilience after the World Trade Center disaster: reconstituting New York City's emergency operations centre. Disasters 27(1):37–53
Kendra J, Wachtendorf T (2006) Improvisation, creativity, and the art of emergency management. Preliminary paper 357, University of Delaware, Disasters Research Center, Newark. http://udspace.udel.edu/handle/19716/3054
Klein N (2007) The shock doctrine: the rise of disaster capitalism. Metropolitan Books, New York
Klijn F, Kreibich H et al (2018) Adaptive flood risk management planning. Mitig Adapt Strat Glob Change 23(7):1087–1107
Lindblom CE, Cohen DK (1979) Usable knowledge: social science and social problem solving. Yale University Press, New Haven, CT
Linkov I, Eisenberg DA, Plourde K, Seager T, Allen JH, Kott A (2014) Resilience metrics for cyber systems. Environ Syst Decis 34:527–530
Lohmann L (2024) Provincializing energy transitions. J Polit Ecol Grass Roots 31:1–1

Manyena B, O'Brien G, O'Keefe P, Rose J (2011) Disaster resilience: a bounce back or bounce forward ability? Local Environ Int J Justice Sustain 16(5):417–424

Matin N, Forrester J, Ensor J (2018) What is equitable resilience? World Dev 109:197–205

Mattern S (2018) Code and clay, dirt and data: 5000 years of urban media. University of Minnesota Press

Matyas D, Pelling M (2015) Positioning resilience for 2015: the role of resistance, incremental adjustment and transformation in disaster risk management policy. Disasters 39:s1–s18. https://doi.org/10.1111/disa.12107

Meerow S, Newell JP (2019) Urban resilience for whom what when where and why? Urban Geography 40(3) 309–329 https://doi.org/10.1080/02723638.2016.1206395

Mbembe A, Nuttall S (2004) Writing the world from an African metropolis. Publ Cult 16:347–372

Moore JW (2015) Capitalism in the web of life: ecology and the accumulation of capital. Verso Books, London

Müller F (2024) Energy colonialism. J Polit Ecol Grass Roots 31(1–16):1–18

Ortiz C, Gómez Córdoba O (2024) Territorial healing: A spatial spiral weaving transformative reparation Planning Theory 23(2) 110–130 https://doi.org/10.1177/14730952231181129

OXFAM (2024) Locally led emergency response fund (LL-ERF) project: experiences from Bangladesh, Puerto Rico, and Uganda. Report, Nov 2024. https://policy-practice.oxfam.org/resources/locally-led-emergency-response-fund-ll-erf-project-experiences-from-bangladesh-621585/

Pain R (2021) Geotrauma: violence, place and repossession. Prog Hum Geogr 45(5):972–989

Papadopoulos D, Puig de la Bellacasa M, Tacchetti M (eds) (2023) Ecological reparation. Repair, remediation and resurgence in social and environmental conflict. Bristol University Press, Bristol

Peck J, Tickell A (2002) Neoliberalizing Space Antipode 34(3) 380–404 10.1111/anti.2002.34.issue-3 https://doi.org/10.1111/1467-8330.00247

Peek L, Tobin J, Adams RM (2020) A framework for convergence research in the hazards and disaster field: the natural hazards engineering research infrastructure CONVERGE facility. Front Built Environ 6:110

Pelling M, Dill K (2010) Disaster politics: tipping points for change in the adaptation of sociopolitical regimes. Prog Hum Geogr 34(1):21–37

Pescaroli G, Alexander D (2018) Understanding compound, interconnected, interacting, and cascading risks: a holistic framework. Risk Anal 38(11):2245–2257

Pickerill J, Chatterton P (2006) Notes towards autonomous geographies: creation, resistance and self-management as survival tactics. Prog Hum Geogr 30(6):30–746

Pinchin K (2019) How Hurricane Maria fueled Puerto Rico's resistance. PBS Frontline, 2 Aug 2019. https://www.pbs.org/wgbh/frontline/article/how-hurricane-maria-fueled-puerto-ricos-resistance/

Puig de la Bellacasa M (2011) Matters of care in technoscience: assembling neglected things. Soc Stud Sci 41(1):85–106

Puig de la Bellacasa M (2012) 'Nothing comes without its world': thinking with care. Sociol Rev 60(2):197–216

Puig de la Bellacasa M (2017) Matters of care: speculative ethics in more than human worlds. University of Minnesota Press, Minneapolis

Puig de la Bellacasa M (2022) Embracing breakdown. In: Papadopoulos D, Puig de la Bellacasa M, Myers N (eds) Reactivating elements. Chemistry, ecology, practice. Duke University Press, Durham, NC, pp 195–230

Quarantelli EL (2000) Emergencies, disasters and catastrophes are different phenomena. Preliminary paper #304. Disaster Research Center, University of Delaware, Newark, DE

Ravazzi S (2024) Improvisation and emergency management: policy responses to COVID-19 in Italy. Palgrave Macmillan, Cham

Rodríguez H, Trainor J, Quarantelli EL (2006) Rising to the challenges of a catastrophe: the emergent and prosocial behavior behaviour following Hurricane Katrina. Ann Am Acad Polit Soc Sci 604(1):82–101

Roux-Dufort C, Vidaillet B (2003) The difficulties of improvising in a crisis situation—a case study. Int Stud Manag Organ 33(1):86–115. https://doi.org/10.1080/00208825.2003.11043675

Ruwanpura KN, Cederlöf G, Ramasar V (2025) Decolonising polycrisis: southern perspectives on interlocking crises. Environ Plan A Econ Space. https://doi.org/10.1177/0308518X251345308

Schuller M, Maldonado JK (2016) Disaster capitalism. Ann Anthropol Pract 40(1):61–72. https://doi.org/10.1111/napa.12088

Shapiro J, McNeish J-A (eds) (2021) Our extractive age: expressions of violence and resistance. Routledge, New York

Simone A (2013) Cities of uncertainty: Jakarta, the urban majority, and inventive political technologies. Theory Cult Soc 30(7–8):243–263

Simone A (2019) Improvised lives: rhythms of endurance in an urban south. Polity Press, Cambridge

Simone A (2022) The surrounds: urban life within and beyond capture. Duke University Press, Durham

Simone A, Pieterse E (2018) New urban worlds: inhabiting dissonant times. Wiley, Hoboken

Singh H, Srivastava SK (2024) From firestick to satellites: technological advancement and indigenous cultural practice in managing forest fires in Australia. Hist Environ Policy Pract 1–24. https://doi.org/10.1080/17567505.2024.2425246

Solnit R (2009) A paradise built in hell: the extraordinary communities that arise in disaster. Penguin Books, London

Soto-Hernández D (2025) Caring about energy transformations: lessons from the Atacama Desert. Hum Geogr 18(2):146–161

Sultana F (2022) The unbearable heaviness of climate coloniality. Polit Geogr 99:102638

Tierney K (2019) Disasters: a sociological approach. Polity Press, Cambridge

Tlostanova MV, Mignolo WD (2012) Learning to unlearn: decolonial reflections from Eurasia and the Americas. The Ohio State University Press, Columbus

Tooze A (2021) Shutdown: how COVID shook the world's economy. Viking, London

Tsing LA (2015) The mushroom at the end of the world: on the possibility of life in capitalist ruins. Princeton University Press, Princeton, NJ

Tsing LA, Swanson H, Gan E, Bubandt N (eds) (2017) Arts of living on a damaged planet. University of Minnesota Press, Minneapolis and London

Vázquez-Varela C, Martínez-Navarro JM, Abad-González L (2022) Traditional fire knowledge: a thematic synthesis approach. Fire 5(2):47. https://doi.org/10.3390/fire5020047

Wachtendorf T, Kendra JM, Butchko K (2015) Disaster improvisation and emergent structures: response to the World Trade Center disaster. Int J Mass Emerg Disasters 33(3):281–309

Wahl DC (2016) Designing regenerative cultures. Triarchy Press, Portland

Wright EO (2013) Transforming capitalism through real utopias. Ir J Sociol 21(2):6–40

Wynn Kirby P (2023) Radiant scars: fallout trauma ghosts and (re)worlding in Fukushima Geografiska Annaler: Series B Human Geography 105(3) 211–227 https://doi.org/10.1080/04353684.2022.2101136

Zaragocin S, Caretta MA (2020) Cuerpo-territorio: a decolonial feminist geographical method for the study of embodiment. Ann Am Assoc Geogr. https://doi.org/10.1080/24694452.2020.1812370

MIX
Papier aus verantwortungsvollen Quellen
Paper from responsible sources
FSC® C105338

If you have any concerns about our products,
you can contact us on
ProductSafety@springernature.com

In case Publisher is established outside the EU,
the EU authorized representative is:
Springer Nature Customer Service Center GmbH
Europaplatz 3, 69115 Heidelberg, Germany

Printed by Libri Plureos GmbH
in Hamburg, Germany